SCIENTIFIC

PAPERS

and

PRESENTATIONS

Communication, like the Celtic tree of life, is a living thing rooted in numerous influences

SCIENTIFIC
PAPERS
and
PRESENTATIONS

REVISED EDITION

MARTHA DAVIS

Department of Crop, Soil, and Environmental Sciences
Dale Bumpers College of Agricultural, Food, and Life Sciences
University of Arkansas
Fayetteville, Arkansas

ILLUSTRATIONS BY
GLORIA FRY

COVER AND FRONTISPIECE BY
SHERI WHEELER WILTSE

ACADEMIC PRESS
An imprint of Elsevier

San Diego   San Francisco   New York   Boston   London   Sydney   Tokyo

ACADEMIC PRESS

An imprint of Elsevier

200 Wheeler Road, 6th Floor, Burlington, Massachusetts 01803, USA
525 B Street, Suite 1900, San Diego, California 92101-4495, USA
84 Theobald's Road, London WC1X 8RR, UK

This book is printed on acid-free paper. ∞

Library of Congress Cataloging-in-Publication Data

Davis, Martha, 1935-
 Scientific papers and presentations / Martha Davis ; illustrations by Gloria Fry. – 2nd ed.
  p. cm.
 Includes bibliographical references and index.
 ISBN 0-12-088424-0
  1. Technical writing. 2. Communication of technical information. I. Title

 T11.D324 2004
 501′.4–dc22   2004050500

British Library Cataloguing in Publication Data
A catalogue record for this book is available from the British Library

ISBN: 0-12-088424-0

For all information on all Academic Press publications
visit our Web site at www.academicpress.com

Printed in the United States of America

04   05   06   07   08   09   9   8   7   6   5   4   3   2   1

To
Aaron Davis
Marion Davis Dunagan
Jody Davis

Contents

Preface to the Revised Edition   xiv

Preface to the First Edition   xvii

1

THE SEMANTIC ENVIRONMENT OF SCIENCE

References   9

2

BEFORE YOU BEGIN

Kinds of Scientific Communication   13
Sources of Help   13
Pencils, Mice, and Cyberspace   15
References   18

3

ORGANIZING AND WRITING A ROUGH DRAFT

Thinking and Writing   22
Prewriting Exercises   22
Organization and Development   26
Coordinating Organization and Development   28
Organizing and Your Point of Emphasis   29
Transitions   30
Writing the Rough Draft   31
References   34

4

SEARCHING AND REVIEWING SCIENTIFIC LITERATURE

Planning the Literature Search   35
Finding the Literature   39
Selecting and Evaluating the Literature   42
The Literature Review   46
References   49

5

THE PROPOSAL

The Graduate Proposal   51
The Grant Proposal   52
Content and Form   53
Other Considerations   63
Progress Reports   64
References   65

6

THE GRADUATE THESIS

The Thesis and Your Graduate Program   67
Avoiding Problems   70
Planning the Thesis   76
Acceptable Forms for Theses   77
The Thesis Defense   82
References   83

7

PUBLISHING IN SCIENTIFIC JOURNALS

Planning and Writing the Paper   86
After the Paper Is Written   87

The Editing and Reviewing Process   89
References   92

8

STYLE AND ACCURACY IN THE FINAL DRAFT

Style   93
Styles in Headings   95
Accuracy and Style in Documentation   96
Proofreading   99
References   101

9

REVIEWING AND REVISING

Reviewing and Revising Your Own Paper   104
Making Use of Reviewers' Suggestions   106
Reviewing the Work of Others   107
References   109

10

TITLES AND ABSTRACTS

Titles   111
Abstracts   112
References   114

11

PRESENTING DATA

Tables   117
Figures   120
References   123

12

ETHICAL AND LEGAL ISSUES

Ethics in Scientific Communication   125

Professional Respect for Others   128

The Legal Issues: Copyrights and Patents   133

References   137

13

SCIENTIFIC PRESENTATIONS

Departmental Seminars   140

The Professional Meeting   141

Speaking at the Job Interview   144

The Question and Answer Session   146

Role of the Moderator   148

Fit the Occasion   149

References   150

14

COMMUNICATION WITHOUT WORDS

Symbols   152

Type Styles   153

Color   154

Physical Communication   156

Listening   159

References   161

15

VISUAL AIDS TO COMMUNICATION

Slide Composition   165

Slide Production   172

References   173

16

THE ORAL PRESENTATION

Conditioning Yourself 176
Timing 178
The Visual Aids 179
Coordinating the Visual Aids and the Speech 179
Transitions in a Slide Presentation 183
The Peer Review 187
Check List for Scientific Slide Presentation 187
References 189

17

POSTER PRESENTATIONS

Audience 192
Text 194
Type Size and Style 197
Color and Physical Qualities 197
Spacing and Arrangement 199
Presentation of Data 200
The Presenter 201
Handouts 202
Making It Fit 202
Time and Construction 203
References 204

18

GROUP COMMUNICATIONS

Group Communication with No Audience 205
Group Communication with an Audience 209
References 214

19

COMMUNICATING WITH OTHER AUDIENCES

Audiences   216
Avenues   219
Subject   220
Techniques   221
Science Writing and Presentation   222
References   223

20

TO THE INTERNATIONAL STUDENT

Becoming Adjusted to U.S. Cultures   226
General Cultural Differences   228
Scientific Writing in American English   230
Oral Presentations   234
Casual Conversation   236
Body Language   237
References   240

Appendix 1
Weaknesses in Scientific Writing   243

Appendix 2
The First Draft   249

Appendix 3
Sample Manuscript   253

Appendix 4
Sample Literature Review   261

Appendix 5
Sample Proposal   269

Appendix 6
Alternate Routes to the Thesis   279

Appendix 7
Sample Review of Manuscript Submitted for Publication   285

Appendix 8
Evolution of a Title   289

Appendix 9
Evolution of an Abstract   293

Appendix 10
Putting Data into Tables and Figures   297

Appendix 11
Sample Letter Requesting Copyright Permission   303

Appendix 12
Use of Color in Visual Aids   305

Appendix 13
Sample Slides and Slide Set   311

Appendix 14
Oral Presentations at Meetings   323

Appendix 15
Sample Text for Poster   329

Annotated Bibliography of Select References   335

Index   341

Preface to
the Revised Edition

The Celtic culture was based upon a complex, interwoven set of beliefs as depicted in the characteristic representation of their tree of life on the cover and in the frontispiece of this book. The Celts were a powerful but somewhat disjunctive civilization bound together by common convictions. For the most part, they have become recognized chiefly for an influential cultural and linguistic legacy that is still quite evident in Great Britain and Western Europe. Their influence on succeeding civilizations, cultures, and languages could have rivaled that of the Greeks and Romans in the Western world with one major exception: they essentially had no written language whereby to preserve and transfer information to those who followed. Instead, they depended almost wholly on oral communications to transmit their knowledge of law, science, history, and culture from generation to generation. Consequently, what we know of the Celts comes to us through the writings of other cultures. Their visual communications were exceptionally strong and remain with us today in the Celtic art and statuary found throughout Europe. Through visual imagery, the Celts preserved some knowledge of their culture and symbols for today's world. Their images reflect the fractals in nature and are now used as symbols communicating a wide range of ideas from love to the continuing cycle of life.

Preservation of the legendary Celtic oral tradition did not fare as well as their art, except when stories were recorded by early Christian scholars. The ultimate flaw in the Celtic culture was in their system of communications. A nonexistent written tradition and over-reliance on oral communications leave us little knowledge of an ancient and once powerful civilization. They have provided a compelling demonstration of the importance of both writing and speaking, especially in science where an immediate or oral presentation of information and a more lasting written form of knowledge are needed to sustain the progress of learning. As Scott Montgomery suggests, "Science exists because scientists are writers and speakers."

The Celtic tree of life, as designed here by Sheri Wheeler Wiltse, can represent a view of life, but also it is a fitting image of the complexity of communication itself with its evolving languages and various roots and branches, visual images, and the myriad of cultural and environmental influences in the semantic ecosystem. In spite of the impossibility of understanding all the components that evolve, grow, interact and interweave to make up the tree of semantics, our job as communicators of scientific information is to simplify this complexity as much as possible for other scientists as well as non-scientists whose work and lives are influenced by our messages. The goal of this second edition of *Scientific Papers and Presentations* is to improve upon the earlier attempt to aid graduate students and other scientists in their efforts to understand what influences their scientific papers and speeches and to communicate more clearly both with the written and spoken word and visual imagery.

This edition has been extensively revised, expanded, and updated. Like the first edition, most chapters are relatively independent, and you can go directly to the chapter that deals with the subject you are interested in. However, I suggest that you read the first two or three chapters to get an idea of my own attitudes toward communication. Additions and changes to this edition have again been motivated by the questions and needs of my graduate students in the sciences. Probably the questions that I get most often begin with "Where can I find..." or "Can you show me an example of...". I've tried to answer these with a concentrated attempt to update the literature as thoroughly as possible and to add more appendices to provide ready, recent sources or examples that can be helpful for writing, speaking, and using visual aids. I have added a chapter dedicated to international students studying science in the United States. And I have made a fervent effort throughout to improve my own communication with my readers. I will have been successful, albeit less than perfect, if this edition proves valuable to students of science.

I again acknowledge with sincere gratitude those mentioned in the preface to the first edition for their invaluable contributions to this version as well. From among them, let me accentuate for this edition the support of Jody Davis for her reviews and her computer enhancement of the tree of life drawing. I am certain that without Jody I would have given up or still be working on the computer technology that goes into producing a text today. Also, thanks again to Marion Davis Dunagan, this time especially for her enlightenment about the Celts. I reiterate the contribution of Luti Salisbury in updating the material for literature searches. Additions to the list of contributors are equally important. First and foremost is Sheri Wiltse whose idea of drawing the Celtic tree of life gave me a vehicle for expressing the complexity of communication and whose influence and encouragement have become a striking enrichment to my life. I thank Magnolia Ariza-Nieto, Andrea Wilson, and Vibha Sirvastava for use of their poster. For their suggestions relative to the new chapter on international students, I am indebted to the comments of Carol Ojano, Pengyin Chen, Elizabeth Maeda, Luis Maas, Ali Jifri, Maria Mashingo, Nilesh Dighe, Palika Dias, Christian Bomblat, and Wenjun Pan as well as other international students who have enriched my classes.

And always, I want to recognize that major force in supporting all my efforts, Aaron Davis.

MD, 2004

Preface to
the First Edition

Whether it is a chemical structure, the anatomy of a rose, or an odyssean siren hidden in the recesses of a DNA code, something stirs the curiosity and lures people into the realm of science. Most scientists are effective, intelligent, logically thinking individuals who are coordinated enough not to destroy the laboratory or the field plans and samples, but many of them become frustrated with communication. The siren did not tell them that many hours of their scientific days would be spent writing reports, preparing for presentations, serving with committees to solve problems, or telling the non-scientist about the value of the science. This book is an attempt to alleviate some of those frustrations with papers and presentations.

Because it is a single, relatively brief volume, I cannot hope to treat every kind of scientific communication in great detail. Other more definitive books concentrate on their respective subjects such as writing skills, journal article publication, writing proposals, group communications, public speaking, and all the other topics to which I have dedicated single chapters. My purpose here is to introduce fledgling scientists to most of the kinds of professional communication that will confront them during graduate studies and as career scientists. My objectives are (1) to answer the basic questions that might be asked about scientific communications and (2) to refer the scientist to more detailed sources of information.

To accomplish these objectives, the first part of the book proposes some practical ideas relative to preparing for, organizing, and producing a rough draft of any scientific paper or presentation. From these general concepts, the book moves to specific written forms that graduate students in science will likely encounter—the literature review, the research proposal, the graduate thesis, the journal article, and the practices and problems that accompany these forms. In scientific writing and speaking, it is important to understand publication styles, abstracts and titles, presenting data, reviewing and revising, and even ethics, copyrights, and patents.

Although a clear distinction between written and oral forms of scientific communication cannot be drawn, I have concentrated in the latter part of the book on slide presentations, communication without words, effective visuals, poster presentations, and oral group communications. I include a chapter on communicating with non-scientists in both writing and speaking.

The appendices and the references are perhaps more important than they are in most books. The appendices provide additional information and examples on the topics discussed in the text. The references extend the views on communication beyond what I can include in this volume. I can give you an introduction to each topic, but you must go elsewhere for other details. At the end of each chapter are references cited in that chapter. Finally, following the last chapter is an annotated bibliography of select works that I find most valuable. I am grateful for what I have discovered in all these sources and believe reference to them will be valuable to any scientist.

Everyone who begins a career in the sciences would do well to have had courses in technical and scientific writing, public speaking, group communications, graphic design, scientific presentations, journalism, leadership and interpersonal skills, professional ethics, audiovisual principles, rhetoric, and other subjects that develop the practical skills of communications. Because taking all these courses would accrue enough college credits for more than a degree in themselves, it is unreasonable to suppose that scientists will be trained in all these areas. They will have little time for practicing writing and speaking skills beyond the efforts required by their work. Similarly, the reading load for a given specialized area of science does not allow time to read all the books that have been written on the formats and skills used in scientific communication. This single handbook will often answer the questions that a graduate student or scientist would ask about scientific papers and presentations and will provide references that can lead to more comprehensive information.

I owe a debt of gratitude to all the graduate students in the sciences who have asked the questions I attempt to answer here. They have provided the motivation and stimulation that have prompted me to put my lecture notes in a form to provide assistance in communication for any graduate student, scientific neophyte, or even the seasoned scientist who may consult this book. I appreciate all that these students have taught me. A special thank you is in order for Terry Gentry, David Mersky, and Katie Teague who have allowed me to use their work in my examples.

I also appreciate all the colleagues who have encouraged me in my teaching and writing, but especially helpful has been Duane Wolf. Without him I would never have pursued this or many other projects. His contributions to the subject matter and to my morale are immeasurable. Thanks, DCW.

Others have helped with the writing, reviewing, and revising, and I truly appreciate their time and attention. Thanks to Marion Davis, who read the whole thing, scrubbed out much of the wordiness, and complained very little the whole time. Thanks to Jody Davis, who read much of it too and was always there to buffer me from the word processor and to keep me motivated. Special thanks to Nora Ransom who gave helpful suggestions on much of the text. Other reviewers on particular parts include Carole Lane, Sara Gregg, Rick Meyer, Lisa Wood, Lutishoor Salisbury, Bob Brady, Domenic Fuccillo, and Bill DeWeese. Suggestions from all of you have helped tremendously.

For the appendices, Justin Morris provided information for the abstract; Gail Vander Stoep allowed me to use her review of a manuscript; and the editorial, "Let There Be Stoning," is from Jay Lehr. I appreciate all of these contributions and the permission granted by Marilyn Hoch, Senior Editor of *Ground Water*, for the use of Lehr's editorial. I appreciate the permission from Walter de Gruyter & Co. for use of the excerpt from pages 69 to 73 of Eduard Imhof's book, *Cartographic Relief Presentations*. Thanks to E. M. Rutledge, M. A. Gross, and K. E. Earlywine as well as D. C. Wolf for the use of a photograph and the text of their poster.

For artistic contributions, I thank Gloria Fry who has provided most of the illustrations, but also Steve Page for the computer drawing on posters. I appreciate the use of Martha Campbell's cartoon, which appeared originally in the *Phi Delta Kappan* (**73,** 130) and the one from A. (Toos) Grossman, which was published in *The Chronicle of Higher Education* on July 1, 1992.

Finally, a special thanks to Aaron Davis for his constant love and support.

MDavis, 1996

1

THE SEMANTIC
ENVIRONMENT OF
SCIENCE

"If it dies, it's biology, if it blows up, it's chemistry,
if it doesn't work, it's physics."

JOHN WILKES
as quoted from graffiti on a bathroom wall

Scientific communication is essential for helping us use and take care of this earth. Researchers who discover the wonders of science must tell someone about their findings in clear, complete, and concise terms. To add to the pool of scientific knowledge, scientists must synthesize available information with what they discover. If a scientist garbles words or leaves out important points, messages become unclear, and the progress of science suffers.

No special talent is required nor is magic involved in clear scientific communication. It is simply a skill developed for exchanging meaning by use of words and other symbols within a social and scientific environment. Meaning associated with those symbols must be the same for both the sender and the receiver. But either the author or the audience can manipulate the meaning and, being human, both probably will. Communication is the vehicle that carries progress, but it also carries disputes and disruption of progress. Generation gaps, wars, and prejudices result, at least in part, from something communicated.

On the other hand, bridges across generation gaps, peace, and understandings are also results of communication. In scientific communication, be ever wary of the human element, and communicate as concisely, conventionally, and clearly as you can with your audience in mind.

Writing or speaking about scientific research is no more difficult than other things you do. It is rather like building a house. If you have the materials you need and the know-how to put them together, it is just a matter of hard work. The materials come from your own study and research. Any attempt to communicate in science is fruitless without valuable material or content. Once quality ideas and data are available, you put them together with the basic skills of scientific writing or speaking. The hard work is up to you.

In any sort of work, you must learn the names of the tools you use or how to operate the instruments in the processing plant, the lab, the construction site, or the field. You must learn what care has to be taken with equipment and with data, or else you should not be working in that area. Whether it is an ax or an autoclave, equipment can be dangerous, and so can words. Writing or speaking, like chemistry or biology, requires cautious skillful work with the tools available and understanding of the content and premises upon which messages are based.

THE CHRONICLE OF HIGHER EDUCATION   A.J. TOOS

*"Class, who can tell me what Mr. Billingsley did wrong,
in addition to majoring in this discipline?"*

FIGURE 1-1
Periodically remind yourself about the fundamentals of science and about successful communication. (Cartoon from Andrew Toos [July 1, 1992]. *The Chronicle of Higher Education*. Used with author's permission.)

But more so than constructing a house or carrying out scientific experimentation, communication contains much of the human element and is far more subjective than is science and less attentive to empirical data. So to work with communication, you have to recognize that it exudes from the individual human into a social context in which it can become either clarified into meaning or polluted into confusion. That means that what you say or write is modified or tempered by your own personality and beliefs, and its reception depends upon the audience and the other elements in the semantic environment in which you deliver the message.

I will probably use the term "semantic environment" frequently in this book. I picked it up from Neil Postman in his *Crazy Talk, Stupid Talk* (1976), a book I would recommend that everyone read. If you don't find that one, read another of Postman's books; any of them will teach you more about language than I can. I am sure I add my own flavor to the term semantic environment (which Postman indicates he got from George Herbert Mead) to make it more applicable to a discussion of the environment in which the language of science is written and spoken. But as it is one of the terms that I will use here to provide you with suggestions on how better to communicate in science, let me expound on it a bit more from my point of view.

The concept of semantic environment is what makes many of us frustrated when a statement is taken out of context. I agree with Postman in that the environment in which words are spoken is essential to the meaning of those words, and any element in that environment can alter a meaning. I like his analogy of pouring a drop of red ink into a beaker of water; all the water in that beaker becomes tinted with red (Postman, 1976). Let's take that analogy outside the lab to a biological ecosystem. Every organism in that ecosystem is influenced by every other organism as well as all the other chemical and physical matter in that environment. The efficacy of the organism relative to its vigor and proliferation depends on the extent to which it can thrive in that environment. The same context can describe your communication efforts. The success of your communication will depend on how well you respond to the multitude of elements in the semantic environment.

For example, suppose you are to make a presentation at a meeting of scientists familiar with your scientific field. First and foremost, the environment is scientific and carries the frame, tone, attitudes, atmosphere, and purposes of science, which can be clearly distinguished from the semantic environments of such things as religion, business, war, or lovemaking. As Postman suggests in *Crazy Talk, Stupid Talk* (1976), "A semantic environment includes, first of all, people; second, their purpose; third, the general rules of discourse by which such purposes are usually achieved; and fourth, the particular talk actually being used in the situation." The semantic environment for your presentation is also made up of a plethora of other smaller influences: how fluent you are with language, how big the room is, what the temperature in the room is, how many are in your audience and who they are, what is in the mind of each audience member, how well prepared you are, and what kind of equipment you are using, as well as the words or other symbols you choose to express yourself and with what tone the words come from your mouth. Those, of course, are just a few of the multitude of influences in your semantic ecosystem, but the extent to which you can successfully use the influences that will support you, and modify or exhibit resistance to those that deter your efforts, is the extent to which you will be successful.

In the speaking situation, some of your success may depend on turning the thermostat down a bit so that the audience is not sweltering, or shutting the door so the noise pollution from outside does not enter the ecosystem. A noisy late comer into the room may "color" the audience like Postman's red ink, but you may be able to clear the "water" somewhat by ignoring the late comer and attracting audience attention back to you and your message. The semantic environment can be totally destroyed if the late comer yells, "Fire! The building is burning down. Get out." You need not try to preserve the ecosystem at that point but simply go to the nearest exit yourself. Except for such radical bombing of a situation, you have a great deal of control over the environment, and as the central focus in the room, you can preserve the environment or destroy it with the way you handle the various elements in the ecosystem. What I'd like to do with this book is help you make the best possible use of your communication efforts throughout your career in the semantic environments in which you live and work.

For you, as a communicative organism, to successfully survive and develop in this semantic ecosystem depends on your understanding of that environment and the practice of good communication skills. If you don't want to be the spindly little weed among the giant redwoods of science, develop your communication skills as well as your scientific expertise. Developing communication skills requires a combination of mental and physical activity. It requires regular exercise or practice to move toward perfection; for example, you can't simply let someone tell you how to swim, follow those instructions, and win a national title the first time you enter the water. The same is true with writing or speaking; only with continual practice can you develop and maintain the skills you need. Once you feel comfortable with those skills, you will be able to publish and make presentations, and you may even enjoy writing and speaking. At least you will be a healthy organism among your peers in this semantic ecosystem.

You have been in school for many years of your life; you know how to talk and write. You may or may not have had much of the needed practice in scientific writing, but you probably have had all the grammar and rhetoric courses you want. Don't disparage those courses. Basic instruction in language use is a good foundation for writing and speaking so long as you don't let that instruction inhibit your communication. Sloppy grammar, punctuation, and spelling can distract from a scientific message. But I don't presume to instruct you on points of grammar and basic composition, rather I want to help you give clear meaning to content and achieve your purpose in scientific communication in producing, reviewing, evaluating, and revising papers or presentations. Those tasks can be easier as you define your purpose in communicating and develop guidelines that will work for you.

First of all, you need to come to terms with your **purpose.** Why are you writing or speaking about a certain subject? Obviously, several motivations stimulate your communication. Students often say their purpose is simply to fulfill an assignment. Maybe your reason for writing a thesis is to get a degree, or you are writing a journal article or making a speech to get a promotion. Those are certainly good reasons to write and present. But surely purpose goes beyond those goals. A general purpose is the exchange of scientific knowledge; your specific purpose will depend upon your subject and your audience. You may want scientists in plant breeding to know that cotton fiber initiation begins at anthesis. Once you define that purpose, you just need to develop ideas to answer questions that might be asked about that conclusion. The more specifically you define your purpose, the easier your task will be. When you have defined why you are communicating, your next job is to determine how best to convey your message.

Any communication, and especially information exchange between scientists, is a matter of asking and answering questions. Answering the question before it is asked often averts many problems. In scientific communication, asking questions is the foundation for discovery; providing answers to your colleagues and to future generations adds knowledge to knowledge and keeps scientific progress alive and well. The questions "How are you?" or "What's happening?" to "Have we discovered the final quark?" or "How important is preservation of the tailed toad?" form the foundation for communication whether or not they are asked. If someone didn't wonder about answers, science would be in real trouble. As you consider a paper or a speech for your fellow scientists, decide what questions are in their minds and yours and which ones you can and should answer.

All forms of scientific communications have a great deal in common. Variations in content and organization are imposed by the questions from different audiences and the answers you give. An audience of sixth graders will not ask the same questions that scientists in your discipline will ask, but you can cover the same subject for both groups. In communicating about your work as a scientist, content and organization are clearly influenced by scientific methods of inquiry and reflect recognition of a problem, observation, formulation of a hypothesis, experimentation, collection and analysis of data, and the drawing of conclusions. Notice that each of these steps poses a question that your research, and then your communication, seek to answer. What is the problem? What do you observe about it? What do you hypothesize? How do you experiment or explore for a solution? What data will you collect, and how will you test it? What conclusions can be drawn? The content of your scientific paper will involve some or all of these questions no matter who the audience is.

Another major influence on organization and content is the use of conventional techniques in scientific communication. An audience can understand you if you use familiar communication devices that they understand and expect. For example, the IMRAD format is often used to organize results of scientific experiments for journal articles, laboratory reports, or oral presentations. The acronym IMRAD stands for introduction, methods, results, and discussion. In reference to written reports, Silyn-Roberts (2000) adds an A, for abstract, so that the acronym becomes AIMRAD. These sections are the conventional, or the expected, order for many scientific papers and presentations. A few journals alter this formula and present the results and discussion before the methods. They are simply answering the question "What did you find out?" before "How did you find a solution?" That organization does not negate the convention; it just asks the questions in a different order. This conventional pattern will not serve for organizing other documents such as the literature review or a proposal. The IMRAD format is simply a common example of what is conventional or what the reader or listener expects.

Much of the expected we do without realizing we are following conventions. For example, in English the order we give to words within sentences is generally the subject first followed by the verb and then its object. Notice that the first sentence of this paragraph does not follow that pattern. As a result, the sentence sounds a little strange, and it would perhaps be a better one if it were conventionally ordered. Except when you are trying to call attention to a sentence or to emphasize a certain point, you should give readers what they expect. A major purpose in this book is to outline what is conventional for the forms of scientific communications.

The rest is up to you. In addition to the questions from a given audience and the conventions that have evolved in language, your success in communications depends upon knowing who that audience is, knowing your subject and purpose, and recognizing your own abilities and convictions. You must be alert to all elements in the semantic environment in which you communicate, interact with everything that supports your communication, and develop a resistance to or tolerance of any pollution that enters the ecosystem. As you develop your career as a scientist, periodically remind yourself about the fundamentals of that science and about the fundamentals of successful communication. Visualize your audience and consider your subject and your purpose for communicating. What questions will that audience ask and how can you best answer them? What medium will best convey your message? Finally, every individual communicates differently; you need to think about yourself and your own capabilities.

First, think about your **audience.** They are most important to the interpretation and understanding of your scientific message. However hard you try to send a clear message, the completed communication rests with them. You can't control an audience entirely, but since you are initiating the communication effort, you are responsible for presenting information in a way that is easily interpreted and understood.

For most scientific papers and presentations, your audience may be scientists especially interested in your subject. However, you will need to communicate also with other scientists and with lay audiences. For example, you may defend your thesis before scientists with relatively diverse interests. Your grant proposal may be going to an agency that has no scientist on the staff, or you may need to present science to young children. You may be trying to communicate with a publisher or an editor, or you may need to transfer a new scientific discovery to those who can make practical use of it but have little understanding of the science involved. Think in terms of what type of experience and education the members of the audience have and what their motivation is for listening to you. Their attitudes and expertise can determine how you will present your subject. You would not make the same presentation to a group of oncologists that you would make to a group of cancer survivors, even if the subject is the same.

Regardless of their prestige and education, members of the audience are human, and so are you. Human beings are rarely logical, fair, and unemotional. No matter how much you try to keep scientific communication strictly factual and objective, the human element is present. For example, if you are making a speech, the audience will notice your appearance and your voice before they hear a word you say. When readers look at a page, they may notice its appearance: the size of type, whether paragraphs are short or long, whether there are headnotes or illustrations. People have certain expectations about how a speaker should dress and sound and about how words on a page should look.

Once words are introduced, the reading audience or listeners have further expectations about meanings and patterns for those words. Most educated people expect standard English diction and sentence construction. If either is substandard or foreign to them, a break in communication results. Whether you are talking or writing, if you first give the receivers what they expect or what they find familiar, they can feel comfortable. You can then lead them to your point even if it is unexpected or unfamiliar. H. W. Beecher said, "All words are pegs to hang ideas on." It's not always the meaning of words that matters as much as it is the way we hang them together and the semantic environment in which they are presented. No word can be fully defined except with the context in which it is sent and how the audience receives it. **The extent to which a word or idea reaches the audience with the same meaning it had when it left the sender constitutes clarity in communication.** Success requires simplicity.

Other things to consider before you begin to write or speak are your **subject** and **purpose** in relationship to the audience. You have to be convinced that your subject is worthwhile, that both you and the audience care about it, and that what you are telling is accurate. For your own confidence, you need a clear purpose. *Conviction* is a key word that joins with *concise* and *conventional* in producing clear communication. (God bless alliteration.) After you are convinced, the next job is to convince your audience. The same subject can be addressed to a group of scientific specialists or to third-grade science students, but not with the same words and techniques. The form your subject takes and the purpose you pursue will partly be determined by who the audience is. A main reason for always having someone review your work is to determine whether your subject has been adjusted to your audience.

With your audience, subject, and purpose in mind, remember that communication is essentially a question and answer format, but often the question is not written or voiced. Your communication will likely succeed if you are answering the same question that is being asked about your subject. That job may sound simple, but be judicious in deciding what questions to answer. For most scientific papers, you answer the basic questions. You will tell what you did and why you did it in the Introduction, how you did it in the Methods section, what you found out in the Results, and what it all means in the Discussion. But it is difficult to get into the minds of your audience, and all their little questions can be frustrating. Who are you? Why did you do it that way? Is that like the result that Jones got? Can you give me an example? These questions are not always the obvious ones that would be asked, but they require answers if minds in the audience are asking. Make your best effort to determine what questions the audience would want answered.

In addition to thinking about the audience and the subject and which questions they would ask about that subject, you also need to think about **yourself.** Attitude is as important in scientific communication as it is in all our activities and accomplishments. If you hate to write or speak, you probably will not do either well. If you love to communicate and find everything you write or speak a true delight, you also likely will not communicate well. A scientific attitude of confidence but of careful self-scrutiny and a dash of humility can form the foundation you need for successful papers and presentations. Allow yourself to be creative, but keep an element of scientific control on your compositions.

You have to believe in what you are doing and what you are saying. Intelligence, education, and personality go a long way, but when you get down to the nitty-gritty of doing science and reporting it, you have to have conviction. This dedication can carry you far in science and always shows in your communication. Don't misjudge your audience. They can tell when you are bluffing, when you don't care, and when you don't believe in what you are doing or saying. You can learn rules for communication and you may even be good at play acting, but without some ability, integrity, and sincerity, your efforts in communication will fall far short of excellent.

Notice how all the elements in communication are interwoven. I cannot talk about subject without discussing audience and purpose and author. Conviction and convention also depend upon all those things. In fact, we can understand any one of the major elements in communication only in relation to all the others. This ecosystem is interlaced with many codependent forces. That is what the semantic environment is all about. There are no rules, but there are conventions. There are questions to ask and answers to provide. An interaction of author and audience with subject and purpose through technique produces communication. In this complex of influences, develop the skills to keep it as simple as possible.

This handbook surveys the forms of scientific communication that a graduate student or a scientist will most likely encounter, and it suggests approaches to the most common problems associated with developing necessary skills. It also gives you other sources to consult. But all the instructions and good examples in the world cannot make you a good writer or speaker. A handbook may be able to tell you how to swim, but you have to get in the water and develop the skill and keep your muscles toned for future swimming. Above all, know how to keep your head above water or be able to come up for air when you need it. Your best source of information on those judgments is you. I hope this text is valuable in helping you to make critical distinctions between what is or is not effective in the scientific exchange of information.

References

Postman, N. (1976). *Crazy Talk, Stupid Talk: How We Defeat Ourselves by the Way We Talk—And What to Do About It*. Dell, New York.

Silyn-Roberts, H. (2000). *Writing for Science and Engineering: Papers, Presentations and Reports*. Butterworth-Heinemann, Oxford.

2

BEFORE YOU BEGIN

"He has half the deed done, who has made a beginning."

HORACE

Before you begin to prepare the paper or presentation, you need to think about your philosophy or attitude toward communication and to determine what kind of communication you are preparing for a particular audience. Determining the kind of communication is rather easy, and you can usually follow patterns or formats that have been developed over the years for organizing a paper or a speech to meet audience expectations. Attitude is another matter. It is ingrained in or developed with your personality and will be different for each individual. But you can condition it with thinking. Keep in mind that this attitude will be a strong influence on how successfully you communicate.

We need a philosophy behind our communication because, as humans, we can generally figure out how to do things (in this situation, write or speak) if we know why. Perhaps our primary struggle as humans is to live, but a close second need is to communicate with others—first with our parents and then with family and friends and with teachers and peers, but also as a scientist with other scientists. Along the way, we develop vocabularies and conventions for putting words or other symbols together so that we will be understood. It is frustrating and often destructive to say the wrong thing or to be misunderstood.

Enter human nature. Attitude may be the strongest influence on what is communicated and how it is interpreted by an audience. We can hardly change our personalities, but we can adjust our attitudes and philosophies. Bad attitudes in scientists and others are often derived from such feelings as the following: "Be impressed with me; I'm important," "I'm scared; feel sorry for me," and "I have a poor background and will never be a good communicator." The scientist with a good attitude sincerely wants to communicate to an audience with confidence and with a dash of humility. The person with a good attitude simply wants to be honest, objective, and clear in sending and receiving messages. Most scientists I know have this attitude.

It is a responsibility of a speaker or writer to present ideas as clearly as possible, but it is also the responsibility of a listener or reader to make an effort to interpret a message clearly and justly. As babies we mispronounce words. *Wawa* may mean "water" or "I want a drink," and the parent will interpret it accurately but may smile at the mispronunciation. We often laugh at inadequacies or differences in communication as children or persons from other sections of the country or other countries attempt to communicate in our vernacular. But our own colloquial language may be no better than that of others, and they may smile at the way we say things. It is good to be able not to take our linguistic foibles too seriously, but it is destructive to laugh if the person trying to be understood is not laughing too. Some groups will develop their own ways of saying things that are not well understood by others—for example, the street talk of some young people or the jargon of social scientists. These creative developments can add to the language and our ability to express ourselves, but the desire to hide meanings or reserve them for a particular group can be destructive and lead to misunderstanding.

Communication is a vehicle for conveying ideas—a network, similar to a nervous system or life blood system that flows through the body of humanity or the smaller environment of science. It can create or kill. If it flows smoothly, it carries civilization forward. It pools knowledge onto knowledge to afford us a background to find the answers to questions that our grandparents would not have even known to ask. If the flow gets clogged, it can completely disrupt a semantic environment. The sins of language, like other sins, can come with excess, omissions, dishonesty, ignorance, or malicious intent. It is a scientist's responsibility to guard against these forces. Respect for language is a serious matter, and miscommunication is dangerous. Enter here the world of science.

The particular language of science presents some differences from other semantic environments. Art and literature play important roles in our world. They delight our senses, stimulate our minds, and teach us a great deal about each other. But the language we use in the complexities of art and poetic language can be distractions to the clarity of a scientific message. The complexity of communicating in science is compounded by the fact that science must be objective, but communication is interpersonal, psychological, and social as well as objective and intellectual. In communication in science, it is essential that one be as simple and exact as possible. For that reason, usually we need to follow conventions that are set forth, follow the expected formats and techniques, and choose words precisely. Scientists can be interesting and creative in their expressions, but embellishment of language must be used to increase audience understanding of a message not to increase the number of possible interpretations, as is often done in art and creative literature.

Nora Ransom's primary rule (for a complete list, see Chapter 3) for scientific and technical writing is "If it can be interpreted in more than one way, it is wrong." That's a strong statement, and we can't always condition the audience to think exactly as we are thinking when we send a message, but as nearly as possible the scientists sending a message must put forth information in the simplest, most easily understood way, and the receiver of that message must listen and interpret as honestly as possible. The progress of science is at stake. You will become more comfortable with scientific papers and presentations by knowing the conventions and tools for the various kinds of scientific communication.

KINDS OF SCIENTIFIC COMMUNICATION

The most common forms for scientific communications are reports, journal articles, proposals, theses, abstracts, speeches or slide presentations, poster presentations, and sometimes books, chapters, review papers, and group communications. **Report** is really a catch-all term that includes everything from a laboratory account of a single experiment to progress reports and group reports on entire research programs. Your chief interest in graduate school may be the **graduate proposal** and the **thesis** or **dissertation,** but as a scientist, you will also become concerned with producing **grant proposals, journal articles, abstracts, slide presentations,** and **posters.** Acquaint yourself with all these forms and produce them with simplicity, precision, clarity, and always honesty. Your first attempts with scientific papers and presentations may benefit from some help. In addition to this handbook, become familiar with other sources that can save you time and improve your communication.

SOURCES OF HELP

If you are just beginning graduate school, you should immediately acquire any information available through your graduate office or your department on writing theses, proposals, or other reports. Also, read *On Being a Scientist* (Committee on Science, Engineering, and Public Policy, 1995); that's something every young scientist should read. I recommend *Getting What You Came For* (Peters, 1997) for a realistic view of getting over the hurdles in graduate school. Smith (1998) gives students in the sciences a good guide for pursuing their graduate degrees. And Stock's (1985) guide to graduate studies is helpful, not only in talking about communication but also in understanding all the other chores that a graduate program and research require. Finally, for success in graduate school, become familiar right away with the library services and computer data bases for literature searches in your discipline.

If you are just entering graduate school or if you have had problems in writing or speaking before a group, spend a few hours reading about communication. The time spent may save you valuable time later when you are ready to report results of your research to the scientific community. Hundreds of books have been published about writing, and most of them contain good information whether they are written well or not. First, I recommend that you read at least Part 1 of *On Writing Well* (Zinsser, 1998). Zinsser is not a scientist, but good writing is good writing, and his comments on simplicity and clutter are in themselves examples of good writing. Then find a rather recent book on scientific writing. Montgomery (2003) would be a good place to start; other good examples I've run across include those by McMillan (2001) and Gilpin and Patchet-Golubev (2000). These books can be valuable at both the advanced undergraduate and the graduate levels, and both contain some answers to stylistic and grammatical questions. You might find more detailed information in books by Silyn-Roberts (2000) or Knisely (2002). O'Connor (1991) also has a great deal to offer about scientific writing. As you get ready to publish and to go to scientific meetings, read Booth (1993) for practical instructions about both writing and speaking. Anholt (1994) offers good information on scientific presentations. Briscoe (1996) and Woolsey (1989) provides good information on poster presentations. All of these sources can serve you as references and give you ideas for developing your attitude about scientific communication.

You won't want to use them for bedtime reading, but you should be familiar with other good reference books and have them handy for your use. Your word processing program will probably have a spell-check, a grammar-check, and a thesaurus. Use them, but they will not always answer your questions. Have your own hard copy of a dictionary, a thesaurus, and a good grammar handbook. The composition handbook you used as a freshman may be good, or check one by Tichy and Fourdrinier (1988). They are thorough about organization, development, grammar, literary style, and diction. For less laborious reading, get *Write Right* by Venolia (1995). It covers the essentials of grammar and punctuation, and the interspersed examples are delightful quotes from others.

In addition, you will find a book on technical writing can be helpful for many of your professional communications, especially short communications such as memos, short reports, letters, and resumes, which I hardly mention in this text. I like the one by Burnett (1994) called *Technical Communication*. Day's *Scientific English* (1995) may be a good choice for stylistic and grammatical answers, and *A Handbook for Scholars* by van Leunen (1992) offers very detailed information. For writing and submitting a manuscript to a scientific journal, Day (1998) offers a cookbook of brief directives and Luellen (2001) provides practical information on writing and publishing. If you need a guide to sources of information in a specific biological discipline, take a look at the guide by Schmidt *et al.* (2002).

Before you begin to design your first scientific illustration, consult Briscoe (1996) or the Council of Biology Editors' (now called the Council of Science Editors) standards in *Illustrating Science* (CBE, 1988). Either of those books will answer far more questions on graphs and other illustrations than I can. Before and again after you look at those books, study data presentation in tables and figures in two or three journals published in your discipline. Not all of these are done well, but you can observe what is acceptable for publication. Those journal articles can also give you a view of their scientific and technical style.

Style, as presented in English composition handbooks, is typically literary style and may not correspond with the technical or editorial style in your scientific area. Learn about the technical style manuals or style sheets that you need to use. *Scientific Style and Format* (CBE, 1994) is a basis for style in almost all of the sciences, and it contains a list of style manuals for special disciplines. *The ACS Style Guide* (Dodd, 1997) goes beyond the details of technical style for chemistry to issues related to publishing (such as copyright, peer reviewing, grammar and punctuation, and ethical guides) and other kinds of communications (such as the oral presentation). You may need to communicate with specific publishers for style sheets peculiar to their publications. Government documents, for example, have unique requirements. If you plan to publish in a journal, check the format and style of the publisher before you begin to write. When writing for journal publication, have handy any instructions from your publisher and a copy or two of the journal to which you are submitting a manuscript. Often it is easier to see how an element of style is handled in the journal or book than to find a rule for it.

PENCILS, MICE, AND CYBERSPACE

In addition to keeping up to date with sources of information, you need to keep current with changes in the language as well as the techniques and electronic equipment available for retrieving information, composing your own paper or presentation, and disseminating your message to others. Language is your primary tool in communication, but other equipment helps you to transfer language from your mind to the minds in your audience.

English teachers have always had a problem with change. It is almost impossible to keep up with what is considered growth and development of language by some or desecration of the mother tongue by others. Whether we like it or not, language is going to change from generation to generation, and as use of a term or construction becomes prevalent, little is gained by objecting to its acceptance. Be alert to all that is new, but I encourage you to err on the conservative side. As Alexander Pope recommended: "Be not the first by whom the new are try'd,/Nor yet the last to lay the old aside."

Particularly important is knowledge about electronic communication and other technologies that can serve your papers and presentations. The computer age has brought with it new vocabulary. There are bits and bytes and CDs and icons. Acronyms become words; permuted words become words. Nouns become verbs. "Input" used to be just a noun; it's a verb now whether or not your dictionary says so. "Online" is an adjective or adverb that describes a condition of the human being; "I'm online now" is perhaps more common than "I'm here." "Fax" is not even listed as a word in my 1981 dictionary, but today it is a noun, an adjective, and a verb with a complete conjugation. It's the English teacher's nightmare. Nonetheless, the computer has given us advancement in communication that is perhaps as great as the invention of the printing press gave to the sixteenth century.

The effects of technology on professional communication are evident. We can send and receive information at a phenomenal rate. Scientists are just a phone call or a computer away from being able to share data immediately. Literature searches are no longer limited by the time one has to plow through hard-copy indexes or by the holdings of one library. The revising and editing of manuscripts no longer require scissors and paste. All of these developments can have a truly positive impact on communication, but we have to look at problems that can result along with the progress. You might want to read *Why Things Bite Back* by Edward Tenner (1996), at least the chapter on computers.

We need to keep a realistic perspective relative to appropriate technology. All the hardware and software in the world will not produce quality research, write the paper for you, or make the successful presentation to an audience.

FIGURE 2-1
All the hardware and software in the world will not write the paper for you.

We must have a human mind, a human hand, and sometimes a human face with the machine. Look at it from the same perspective our forebears must have had when the free-flowing pen and the pencil replaced the quill. Scribes must have been amazed at what the printing press could do for their work in the late fifteenth century. For them, moveable type would have been ingenious, and photocopy would be a miracle. Despite our advances in communication, none of these tools alone will initiate the communication or give to it an accurate interpretation. Don't lose sight of the importance of input and reception by human beings.

Take advantage of available technology, and if you are weak in computer literacy, take advantage of classes, workshops, or sources such as the book by Munger and Campbell (2001). As a scientist you will be handicapped without access to the Internet, to e-mail, and to a fax machine. Learn to compose at the computer. It erases easily and does not leave smudges. Learn to revise and edit from the screen. With data being easier to acquire and analyze, the dangers increase for trying to present too much in a small space or a short time, for being too wordy, or for putting too much information on a single visual aid. Make hard copy to proofread after you have proofread from the screen. Use the grammar- and spell-checks, but recognize their limitations. The spell-check will not put the *y* on *the* to make *they*, nor will it retrieve a word that has been omitted. You may have the problem of the young scientist who presented his research in the poster format at a national meeting. He meant for his objective to begin with "To assess..." but it actually began with "To asses...." He had spell-checked his text; the audience at his poster smiled and were distracted from his scientific message. One audience member even suggested changing the spelling of *to* to *two*.

In addition to composing your manuscript via word processing, publishing from computer discs or simply sending electronic copy through cyberspace has several advantages. Traditionally, hard copy submitted to a publisher had to be reset into type for publication. This process provided opportunity for more errors than are made if material is published without resetting. But electronic publishing is not without possible errors; both the author and the publisher should coordinate their efforts and take care in the transfer and publishing of information.

For your own use, desk-top publishing equipment and software programs can produce text of high quality. Digital cameras, film recorders, and scanners can be connected with computer units to produce slides or prints and posters. Letter-quality and laser printers produce copy that is ideal for posters or slide copy, and many printers now accommodate a large paper size that will produce an entire poster on one sheet. With interactive media you can produce presentations that combine visuals, words, voices, and video action. Find what will work best for your communication effort.

Be alert for possible problems. Backup your input regularly on additional discs and hard drives. Keep hard copies, and don't throw away your pencil. Paper and pencil are not nearly so prone to technical disaster such as losses due to electrical surges or illnesses from viral infections. No equipment is without fault. When you attach people to imperfect equipment, problems can arise. The power of information exchange can be used in unethical ways. Copying software and invading privacy through stolen access codes certainly have occurred. Question the credibility of everything you read whether it is in hard copy or on a computer screen. And as Henderson (2002) says, "When in doubt, doubt." Misinformation, faulty information, or misuse of information are dangers in electronic communication just as they are in books and other hard copy. Several sources are available with suggestions for evaluating material you find. Start with Henderson (2002); his Web site can connect you to other links.

With all the advancements in technology, the tools for communication are decidedly improved, but the basic principles remain the same as they were when the scribe carefully copied the wisdom of Pythagoras with a quill to preserve it and communicate with new generations. Make full use of the keyboard, the mouse, and networks, but don't simply default to what a program wants to do. Good software will allow you to make decisions yourself. Do so, and don't lose sight of what the Chinese and Greeks already knew when history was a child: **Simplicity and clarity are essential in conveying a message.**

Most important of all—don't let the technology dictate to you what constitutes good communication. We should not accept graphs that are too complex simply because the computer has the capability of producing them. We should not accept distortions simply because your computer program will not produce the expected dimensions or form. The technology should in no way dilute clarity. Study the techniques for clear communication, and then make the equipment work for you.

In this book I will deal with age-old principles of communication and hope that you can apply them to our changing world.

References

Anholt, R. R. H. (1994). *Dazzle 'em with Style: The Art of Oral Scientific Presentation*. W. H. Freeman & Co., New York.

Booth, V. (1993). *Communicating in Science: Writing a Scientific Paper and Speaking at Scientific Meetings*, 2nd ed. Cambridge University Press, Cambridge.

Briscoe, M. H. (1996). *Preparing Scientific Illustrations: A Guide to Better Posters, Presentations, and Publications*, 2nd ed. Springer-Verlag, New York.

Burnett, R. E. (1994). *Technical Communication*, 3rd ed. Wadsworth, Belmont, CA.

Committee on Science, Engineering, and Public Policy (1995). *On Being a Scientist*. National Academy of Sciences, National Academy Press, Washington, DC.

Council of Biology Editors (CBE) (1988). *Illustrating Science: Standards for Publication*. CBE, Bethesda, MD.

CBE (1994). *Scientific Style and Format: The CBE Manual for Authors, Editors, and Publishers*, 6th ed. Cambridge University Press, Cambridge.

Day, R. A. (1995). *Scientific English: A Guide for Scientists and Other Professionals*, 2nd ed. Oryx Press, Phoenix, AZ.

Day, R. A. (1998). *How to Write and Publish a Scientific Paper*, 5th ed. Oryx Press, Phoenix, AZ.

Dodd, J. S., ed. (1997). *The ACS Style Guide: A Manual for Authors and Editors*, 2nd ed. American Chemical Society, Washington, DC.

Gilpin, A. A., and Patchet-Golubev, P. (2000). *A Guide to Writing in the Sciences*. University of Toronto Press, Toronto.

Henderson, J. (2002). "ICYousee: T is for Thinking: A Guide to Critical Thinking about What You See on the Web." www.ithaca.edu/library/Training/hott.html (verified July 11, 2003).

Knisely, K. (2002). *A Student Handbook for Writing in Biology*. W. H. Freeman & Co., Gordonsville, VA.

Luellen, W. R. (2001). *Fine-Tuning Your Writing*. Wise Owl, Madison, WI.

McMillan, V. E. (2001). *Writing Papers in the Biological Sciences*, 3rd ed. Bedford/St. Martin's, Boston.

Montgomery, S. L. (2003). *The Chicago Guide to Communicating Science*. University of Chicago Press, Chicago.

Munger, D., and Campbell, S. (2001). *Researching Online*, 4th ed. Longman, New York.

O'Connor, M. (1991). *Writing Successfully in Science*. HarperCollins Academic, London.

Peters, R. L. (1997). *Getting What You Came For: The Smart Student's Guide to Earning a Master's or PhD*, revised edition. The Noonday Press, New York.

Schmidt, D., Davis, E. B., and Jacobs, P. F. (2002). *Using the Biological Literature: A Practical Guide*, 3rd ed. Marcel Dekker, New York.

Silyn-Roberts, H. (2000). *Writing for Science and Engineering: Papers, Presentations and Reports*. Butterworth-Heinemann, Woburn, MA.

Smith, R. V. (1998). *Graduate Research: A Guide for Students in the Sciences*, 3rd ed. University of Washington Press, Seattle.

Stock, M. (1985). *A Practical Guide to Graduate Research*. McGraw-Hill, New York.

Tenner, E. (1996). *Why Things Bite Back: Technology and the Revenge of Unintended Consequences*. Alfred S. Knopf, New York.

Tichy, H. J., and Fourdrinier, S. (1988). *Effective Writing for Engineers, Managers, Scientists*, 2nd ed. John Wiley & Sons, New York.

van Leunen, M-C. (1992). *A Handbook for Scholars*. Oxford University Press, New York.

Venolia, J. (1995). *Write Right: A Desktop Digest of Punctuation, Grammar, and Style*, 3rd ed. Ten Speed Press, Berkeley, CA.

Woolsey, J. D. (1989). Combating poster fatigue: How to use visual grammar and analysis to effect better visual communications. *Trends Neurosci.* **12,** 325–332.

Zinsser, W. (1998). *On Writing Well: The Classic Guide to Writing Non-Fiction*, 6th ed. HarperCollins, New York.

3

ORGANIZING AND
WRITING A ROUGH DRAFT

"...the greatest truths, poorly communicated,
remain unconvincing."

LOIS DEBAKEY

You can do a variety of things before you actually begin to organize and write a paper or put together a speech or a poster. A common choice is to procrastinate. You can even rationalize that the procrastination is leading to better communication. The paper is due soon, and so you take a nap to get a fresh start on it. Maybe you will have a beer with a friend just to relax so that you will do a better job on the paper. You are even willing to wash windows or do laundry to distract yourself and thereby clear your mind for the important paper. The computer is right there, and so you play a computer game; you deserve the little break before the large task before you is begun. Or, no, you won't play, but you can continually revise the list of things you plan to write about. Be creative; you can surely think of many other diversions and rationalize your way into a most artistic, well-crafted procrastination. But at some point, you'll have to give in to the need to produce the paper. For the good student or scientist, self discipline will win despite the numerous skirmishes that procrastination wages in opposition.

Once that battle is won, consider what you may need for the paper or speech at hand. Determine what kind of paper you are to produce, isolate your purpose, and reflect on your subject and your audience. Then set your mind on the paper itself and perhaps go through a few prewriting exercises to condition yourself for producing a rough draft.

THINKING AND WRITING

You think well; you write well. I am convinced that clear writing depends upon clear thinking. And, like Zinsser (1988), I believe writing helps you to think and to learn. Thinking alone allows you to skip around, to move at random through a batch of ideas. The physical symbols (words) you put on paper help to direct your thinking and to make you recognize any gaps in logic and to focus more clearly on your purpose. Some people can visualize physical situations and can work out mental details for such projects as building a barn or repairing a motor. More often, the carpenter will sketch or look at a sketch of the barn, list materials and numbers of board feet needed, or even stand on the building site to visualize the structure. The mechanic will almost always want to look at and listen to the motor. The tangible substances help these workers coordinate their ideas and skills and produce results.

Words on paper have a similar symbolic influence on thinking, learning, and knowing. The physical words keep ideas in perspective and contribute to logical, orderly arrangements of thoughts. However, some people avoid this learning device because they find that moving thought to paper is difficult. Some find the writing itself relatively easy but getting started difficult. Multiple influences go into writing a scientific paper—the research, the data analysis, what others have written, your colleagues' opinions, your audience, your ability with words, the time you have for writing, how tired you are, and what other thoughts are in your mind; the list becomes endless. Juggling your thoughts and having them all fall into place neatly on a page are difficult, but what cannot be done in the initial writing can be done through revision. You must create a rough draft before you can revise.

Students often tell me that getting started is the hard part. Some outline; some don't. Some wait for time to pressure them; some write better without the pressure. Write the way you do it best. But if you are having trouble deciding what is best for you, try some of the following ideas.

PREWRITING EXERCISES

Think Before You Write

Your first efforts should be the thinking about the audience, the subject, and the purpose. Go through a series of questions and see that you are satisfied with your answers. What form will the writing take? How do I talk to this audience?

What are they asking? What are my motivations? Do I believe in what I am saying? Do I have the materials I need, including the literature, the data, the understanding, and the reference books? Jot down brief answers to these questions. Don't linger too long, but think long enough that you don't resort to the excuse of not being ready to write. When you have all the auxiliary matters out of the way and your thinking is clear and a few notes are written down, then go on to another prewriting exercise or start writing the draft.

Talk Before You Write

Prime yourself by trying to tell the material to a colleague, a spouse, a friend, or your pet. You can even talk to yourself via a tape recorder. The tape will hold your thoughts until you need them again. Talking out loud can cause you to recognize the logical progression of ideas or what is missing from your information. Involving your own ears and perhaps those of someone else can help you hear what details need to be included and emphasized and what order is needed for clarity. Request that your listener make comments and ask questions that will show you points that may be clear to you but not to someone else. With or without input from your listener, write immediately after talking while the subject is still up front in your mind.

Brainstorm, Freewrite, or Make a List on Paper

Just start putting words down on paper. This process can consist of a graphic presentation of ideas in boxes or circles joined by arrows, or it can be a running text of sentences as they come to mind. You may want to list the questions you think the audience will want answered. Try note cards; they are easy to shuffle and organize. Once you have a physical list of ideas in one form or another, you will be better able to organize the material. Don't spend a lot of time trying to perfect this list; get to the real draft soon. The point in any kind of brainstorm writing is to get words down on paper. You may throw them all away when you really get organized, but they are the stimulus and the material for your organization.

Outline

You may begin by outlining, or the outline may follow the brainstorming, freewriting, or talking. Some people can use either a rough outline or a perfected one to organize before they write any full sentences. Others may use outlining as an intermediate step to revise a first draft. Write first and then outline and rewrite if that technique works best for you. Once you have produced the outline, don't let it take over. It needs to be flexible so that you can reorganize if need be.

In making a list or a rough outline to work from, you may want to leave the computer screen and take up your trusty pencil and a package of note cards. You can physically arrange and rearrange the note cards, place them all in front of you, and see the first and the last card at the same time. If a computer screen works well for you, use it in the same way, but don't let the mechanical maneuvering of parts distract from the ideas and their connections.

Write a Rough Abstract First

An abstract can be a kind of outlining. The scientific abstract generally requires that you write a sentence of rationale, a statement of objectives, a notation about methods used, a list of most important results, and any conclusion reached. A draft of an abstract will set forth the main points about which the paper must elaborate. Consequently, it can get you started in an organized manner. And, like the outline, it guides you through the entire paper in that you simply expand upon each part of the abstract.

Start in the Middle

Some people have trouble introducing a subject but know what they will say in the main body of the paper. An introduction could be the last part written. Don't waste time by trying to produce a perfect paper from beginning to end at the first sitting. Start in the middle and fill in around the edges as you can. The need for revision is a fact of successful writing. The important thing initially is to get something down on paper to revise.

Get Rid of Your Inhibitions

Loosen up. Most of us have inhibitions about writing, the same as we do about speaking to an audience. The sources of those inhibitions often have come from the same people who intended to teach you to write. Even on papers you wrote for English class, busy teachers probably commented little on the content. Maybe they scrawled an "excellent" or a "good" at the top of your paper. I was also familiar with "very poor." Despite having little to say about your ideas or content, those same teachers' red pens bled freely across the page, marking the mechanical errors in grammar, punctuation, and spelling. And they may have preached the gospel of the Rules. We often reach graduate school still cowering before the mighty Rules. And, alas, the rules are flexible. I have become a heretic. Although I rely on grammar handbooks to show me what is conventional, I don't believe in allowing rules to distract from the content. The best set of rules I have seen proposed for scientific writing was devised by Nora Ransom at Kansas State University. My own comments on major weaknesses in scientific

FIGURE 3-1
We often reach graduate school still cowering before the rules.

writing are in Appendix 1, but the following are the only rules you will find in this book.

Whatever technique you use to get started, get started. Don't spend too much time on these prewriting exercises. Write. Don't let any of the monsters on the fringes of your thoughts deter you. Thoughts of an unreceptive audience, thoughts of other things you could be doing besides writing, hang-ups on mechanics of good usage that your English teacher has instilled in you—all

---

Ransom's Rules for Technical and Scientific Writing

1. If it can be interpreted in more than one way, it's wrong.
2. Know your audience; know your subject; know your purpose.
3. If you can't think of a reason to put a comma in, leave it out.
4. Keep your writing clear, concise, and correct.
5. If it works, do it.

such devils should be driven from your path. The only way to get something written is to write. Do it your way. In your revisions you must be sure your paper or presentation is well organized and that all the language monsters have been subdued so that you are communicating well with your audience. Notice the very rough draft of a simple essay in Appendix 2. It will require extensive revision before it is an acceptable draft, but it was written in 10 minutes and it gives the writer a foundation to revise for a good essay.

ORGANIZATION AND DEVELOPMENT

During the thinking and prewriting processes, you may organize your material within a format. Further organization may, however, be a chore that follows after prewriting exercises or after a freely written rough draft. Organization produces a unified package that makes both sending and receiving information easier. In addition to arrangement of content, you use language tools—words, sentences, and paragraphs as well as symbolic signposts such as transitions or headings and subheadings—to direct the reader or listener through the organization and development of your subject.

Organization and development are concepts that cannot be separated. It may be that a second point in your text will not be understood until the first point is developed. To keep the progression of ideas orderly, keep the audience in mind at all times. As you write, think of someone listening or watching who is asking questions repeatedly: "What did you do" "How did you do it?" "What do you mean?" "What caused that?" "Can you give me an example?" Answers to such questions constitute development of ideas, sentences, and paragraphs. This inquisitive little voice over your shoulder can help you keep things in order and tell you how much development you need. It will remind you when you are using too much jargon or need to explain further or when you are proceeding too fast without making clear the background, the definitions, and the causes and effects.

In writing or reading a paper, we move from word to word, from sentence to sentence, and so on, building larger and larger units of material until the full paper is put together or understood. In organizing a paper, we use the reverse procedure. We first consider the largest units. To an extent, the kind of scientific paper will dictate the largest units. A journal manuscript, for example, will almost invariably follow the IMRAD (Introduction, Methods, Results, and Discussion) format. Proposals may require different framework, possibly Introduction, Justification, Background or Literature Review, Methods, and Conclusions. A thesis may require yet another primary organization, as will a review article. The large sections of your paper may be more lucid if based on chronological order or a step-by-step process; they may be based on spatial arrangements or geographical order; or they may be divided by meanings (two or three comparisons or contrasts, causes and effects, pros and cons).

When the framework is in place, we must consider each smaller and smaller organizational unit. For example, a Methods section may be divided into subsections on materials, steps in experimentation, data collection, and scientific and statistical analysis of data. Under each of these subsections may be tertiary sections that describe treatment 1 and treatment 2 or location 1 and location 2, with smaller divisions under each of these. At some point you can quit organizing and start writing or talking. That point depends on your personality, your expertise, and maybe the way your brain works. Some people start writing as soon as main headings are established; some need details down to the developmental topic for each paragraph or even subtopics within the paragraph. It all sounds very simple, but the important point is to be sure that you extend your organization to the smallest unit you need before you attempt to move in the other direction and write the paper. Notice again that, as you **organize,** you proceed from the large to the small:

Sections
Subsections—Subsections
Sub-subsection—Sub-subsection—Sub-subsection
Paragraphs—Paragraphs—Paragraphs—Paragraphs
Sentences—Sentences—Sentences—Sentences—Sentences
Words—Words—Words—Words—Words—Words—Words—Words

And when you **write,** you go from the smallest to the largest unit:

Words—Words—Words—Words—Words—Words—Words—Words
Sentences—Sentences—Sentences—Sentences—Sentences
Paragraphs—Paragraphs—Paragraphs—Paragraphs
Sub-subsection—Sub-subsection—Sub-subsection
Subsections—Subsections
Sections

Main sections may be labeled with primary headings, subsections with secondary heads, and sub-subsections with tertiary heads. You'll probably not need more specific labeling. Paragraphs indicate new development of an idea, just as capitals and periods tell us where sentences begin and end. I still believe in indenting paragraphs; the indention is just another clue to the audience that you are shifting to a new point for development.

Outlining in your mind and on paper can be the most efficient way of saving yourself time and staying on track. You have likely developed your own methods for outlining or organizing. If you haven't or if your technique doesn't work well, try the following. First, simply write down the big headings—Introduction, Methods, Results, Justification, and so forth—and then start making notes or lists under each heading. You may immediately see an outline emerging. You may, however, need to arrange your list further by subordinating some ideas to others, putting materials into categories, and giving an order to what needs to be told first, second, third, *etc.* Now an outline is emerging. If you are a tidy person, you may want a formal outline. Montgomery (2003) has suggestions for organization. An English handbook can provide guidelines for outlining, or read Chapters 3 and 4 in Tichy and Fourdrinier (1988) on concepts of outlining. An example of a simple outline is in Appendix 3.

COORDINATING ORGANIZATION AND DEVELOPMENT

If you can get basic organization as well as content development into an early draft of your paper, your work with revision will come much easier. Note how organization and development work together in Appendix 2. This very simple essay is a rough draft that could easily have been written by a high school student, but notice the potential for developing such a draft into a bona fide paper. The elements of organization include main points and transitions that carry the reader from one point to another. Now look at Appendix 3. It is still a very simple paper but perhaps illustrates what you need to do with a more complex one. The outline guides the writing from point to point as it follows the typical IMRAD formula, the most common external organization for a scientific report. Notice how the internal organization and the development of content follow standard criteria for each of the parts as described below. For more specific treatment of content for these organizational sections, refer to Day (1998).

The **Introduction** should serve three purposes: (1) to call attention to and clarify or define the specific topic or hypothesis that you are to discuss, (2) to provide background and justify a study relative to its important and the results of other studies, and (3) to list the objectives of your research project or to give the audience information on what you plan to accomplish in your paper.

The **Materials and Methods** section will be a recipe that reveals how you acquired and used your data. Organization here is usually easy, with step-by-step processes kept in the same order as the objectives were listed in the introduction. You can preface this procedure with a listing of materials used, conditions present, or design of the project. In addition to needing details in the materials and methods used, the audience will ask two major questions: Is this researcher's work credible, and can I use the same methods? To answer these questions, you must provide complete information on ingredients, actions, conditions, experimental design, replications, repetitions, and statistical analysis. Ask yourself the following: "Could another scientist follow my words in this part of my paper and perform the same experiment with the same results?" A well-written methods section will support a positive answer.

Results should not keep a scientific reader in suspense. Make clear immediately the extent to which you have proved or disproved your hypothesis and then carry the reader from one display of data to another with logical development, showing how your findings satisfy your objectives. Results may be presented in the same order as the objectives and the experimental procedures. Data are often presented in tables or figures, and the text will simply serve to tie the data to your objectives or to call attention to main points in the data display. In addition to the valid experimental design that you present in your methods, your results and your discussion will establish your credibility.

Discussion is sometimes interwoven with results, or it can be in a separate section. This commentary provides meaning or an interpretation of the results and shows relationships with other research. Summarizing statements will tie together outcomes as depicted by the various data sets. As with results, strongly focus the discussion on your objectives. The discussion should present the overall significance of your work and help direct the thinking of your audience, but once you have made a statement on what your data mean, don't go too far afield with speculation. Show how your results fit into or compare with those from similar studies in the literature, and leave most of the speculation to your readers. Let them form their own opinions. Don't hesitate to suggest a direction you want the receiver's thoughts to take; just don't overdo your own speculation.

Finally, be sure to make concluding statements at the end of your discussion or in a section called **Conclusions.** Briefly reiterate your objectives and provide a general statement on the extent to which you have accomplished them. Be sure you don't just restate the results here, but draw together outcomes of your objectives and enumerate these conclusions succinctly. They may be the points that stay longest in the reader's or listener's mind.

ORGANIZATION AND YOUR POINT OF EMPHASIS

Organization may fall apart unless it is tied together with a thesis. *Thesis*, in this instance, means the theme, the motif, the focus, or the overriding topic around which everything in your paper should revolve. In communication, art and science follow the same principle; a good musical composition or piece of literature carries an overriding motif and lets us hear it or consider it from several vantage points. The question that the scientific audience asks you is "So, what's your point?" Over and over you need to reiterate that point or thesis.

If we use physical imagery to interpret this concept, we can think in terms of the blocks of information that make up the various sections of a research report. A common thread must be woven through these blocks to hold them together. Most scientific papers emphasize one or two main points. The hypothesis is one version of this point and the objective another. Methods provide another view, and results concentrate on what you found out about that point. Any discussion or conclusions must strictly adhere to the point. With concentrated focus on your thesis, the organization is not likely to stray.

I like to think of a well-organized presentation of a clear thesis or a focal point as a common thread, moving in a spiral or circular pattern rather than progressively or linearly from first to last. The first and main point, usually the objective, weaves in and out throughout the paper and comes to rest as a final view of the same point in the conclusion. If you proposed a possible answer to a question (a hypothesis) in the introduction, you conclude with a proposed answer to that question in your conclusion.

TRANSITIONS

When we talk about organization, we are talking about a road map that will carry you and your audience through the various developmental blocks or units of materials that make up your paper. In addition to a thesis or main point of emphasis, we need mortar or bridges to join the blocks, and we need sign posts to direct the reader along the road. If you think of the thesis as a thread, these connections are loops in the thread that sew one section or idea to the next. Those language devices that hold your paper together are **transitions.** They help to unify and move the point of emphasis through a series of ideas to a conclusion. Notice the examples of transitions in Appendix 2.

Transitions serve both to join parts of a subject together and to convey meaning. Like the Roman god Janus, transitions look in both directions at the same time, but they signal movement to the next idea in your paper. They may be conjunctions or prepositions that hold parts of a sentence together with distinctive meanings. For example, the short words *and* and *but* are transitions that carry opposite meanings. The same is true of *to* and *from*. Keep in mind the functions of transition: joining two parts together and carrying the message forward with meaning.

Beyond simple words as transitions, you can use phrases such as "on the other hand" or "in the same way." You can use full sentences: "Chlorine reduced the bacterial populations in the first experiment. In the second, we substituted fluorine." The full sentences are transitional, and they contain the transitional words *first* and *second*. You can also use full paragraphs as transitions between two more complex points. Within those paragraphs, you will likely find several smaller transitional elements.

Another transitional device is repetition. You can repeat the same word or phrase, or you can repeat the same idea in different words. This kind of transition is important for carrying a major point or motif throughout a paper or speech, referring to it again and again to keep the reader or listener focused on that point. Repetition in styles used for headings, in sentence structure, in symbols, or in colors and designs can serve as unifying elements and as transitions. Good transitions make a paper or a speech flow smoothly and keep the communication from being jerky or jumping from one point to the next. For a more detailed but brief discussion on transitional devices, consult Marshek (1982).

Whatever connections you use or however you visualize your unified paper, the overriding requirement for good organization is clear thinking with use of convention and order. In other words, if you do what is expected in a reasonable manner, your writing will be most clearly understood by other reasonable people who speak your language with the same conventions. Whether the break in convention is a misspelled word, an awkward transition, or chaotic organization, it will distract from communication. (For those of you whose first language is not English, you may want to read Chapter 20 and, for writing for an English-speaking audience, work out any differences in organization that your culture has instilled in you.)

WRITING THE ROUGH DRAFT

Once you have corralled your thoughts, with or without prewriting exercises, and considered the organization of your text, you are ready to write a rough draft of a short paper or a section of a longer work. At this stage, I assume that you have focused on one point of emphasis, you have thought through what you want the audience to know when they have finished reading your paper, and you have considered what their questions will be. You should have gone through some prewriting exercises and produced an outline or other organized pattern for your work or have made a list and then organized your ideas so that your audience moves with you from one point to another. With all this preliminary work done, begin to write immediately. Any delay at this point is distracting.

Keep in mind that your first production is a rough draft and will require revision. Perhaps a first draft such as Moses' Ten Commandments could be set in stone; yours should not. Your words are not sacred. When you have written a rough draft, be willing to tear your words apart, throw them away, insert new ones, and rearrange them until they say what you want them to say.

Language is really not so haphazard as it sometimes appears. Lack of clarity results only when a speaker or listener (writer or reader) deviates from logical organization, from standard meanings of words, or from expected form or word order. Think clearly before you speak or write, and the audience should receive your message without misunderstanding.

Even after careful organization, some people have problems getting started. Don't waste time wondering how to start; there are not hundreds of ways to

approach writing. In creative writing, approaches such as stream of consciousness or story-telling can add artistic or philosophic dimension to writing, but for
scientific writing, you will seldom need to know how to handle more than the
following four approaches: **Define, Compare–Contrast, Enumerate, and
Give Cause–Effect.** Don't ponder, and don't use up time at this point. Choose
one (or two) of the four and start to write immediately. The others can be in the
back of your mind and summoned at a moment's notice. Notice the use of these
approaches in Appendix 2.

Compare–Contrast
Enumeration
Definition
Cause–Effect

Numerous key words used in asking and answering the audiences' questions
will control the development in your paper or presentation. You may develop
ideas with evidence, example, detail, points, methods, classification, results,
summary, reasons, alternatives, and possibility. All these words are related to
the four approaches. Evidence may be cause and effect or a definition or an
enumeration of points or a comparison or contrast. The questions from the
audience may take many forms, but the techniques involved in the reply
are limited. Your answer will characteristically begin with a general assertion
followed by details to explain and clarify that assertion.

The audience and their questions will often tell you how to start or, at least, hint
at an approach: "What is a quark?" (definition) or "What caused that result?"
(cause–effect). If you must choose the approach, as in number four below, you
usually know which one you can handle best or which will best serve the audience.

<h2 align="center">EXAMPLES</h2>

1. Question: What are your methods for _____?
Answer: Enumerate 1, 2, 3, 4 methods; then define, probably by enumerating
steps in a process, showing cause and effect, or comparing and contrasting.
2. Question: What are the differences in _____ and _____?
Answer: Define terms and then contrast differences, probably through
enumeration.
3. Question: What is _____, and how does it affect _____?
Answer: Definition and cause–effect.
4. Question: Why should one use continuous rather than intermittent
flooding in rice?
Answer: Here none of the four approaches is obvious, but choose one or two
and begin immediately. Probably you need to enumerate cause and effect:

(1. weed control, 2. less loss of nitrogen, 3. greater yields) and compare–contrast to intermittent flooding (enumerate again).

A common way to begin an essay or a paper is with definition. A safe way (i.e., it will always work) to begin a definition is to put the term into a general classification of related ideas and then point out its distinctive qualities.

Formula: __(term)__ is __(class)__ which __(distinction)__

__term__   __class__

Example: Factorial ANOVA is an analytical technique in

__distinctions__

which two or more variables and their interrelationships can be identi-fied….(Continue to enumerate other distinguishing characteristics.)

Comparison and contrast or cause and effect almost always involve definition and enumeration. Both approaches lead to examples and **tangible details.** A person can get lost in technical terms if tangible images are not occasionally formed in the mind. A tangible image is one that relies on the senses (or a past sensory experience) for understanding. We see, taste, smell, hear, and/or touch tangible examples. Words such as *lemon, freight train, rose, fuzzy*, or *square* throw images into our consciousness to increase our understanding of an idea or concept. As you experience scientific and intellectual concepts, your senses are still at work. Terms such as *hydrogen sulfide* or *staphylococcus* are under-stood in part through your senses of smell or sight once you have encountered the real substance. Notice the difference as we add tangible details in the following:

General	Factorial ANOVA compares relationships of two or more factors.
More specific	Factorial ANOVA was used to determine interrelationships of two treatments on three cultivars at two locations.
Most specific	Factorial ANOVA was used to determine interrelationships between foliar applications of nitrogen and boron on "Forrest," "Davis," and "Clark" soybeans in field plots at Fayetteville and Marianna.

As your communication becomes more specific, you alert your audience to specific images that define and explain your message. No matter how theoreti-cal your scientific communication must be, keep as close to tangible details as possible. If the senses cannot be involved to make the details tangible, at least be as specific as possible with intellectual concepts.

A great deal of time is wasted in struggling with the first rough draft, and focusing your own mind on the tangible details can help. Don't worry with perfection at this point; you simply need a draft to begin working with. If nec-essary, put a time limit on yourself, ask yourself the appropriate questions, and then write details with as much organization as possible. Although the example in

Appendix 2 is hardly scientific, notice how it displays organization, development, transitions, and the four approaches to writing.

As you gain experience in writing, you will use the basic approaches to organization and development of ideas without realizing that you are doing so, just as you use periods at the ends of sentences and start the next with a capital without having to remind yourself each time. But when you are having trouble getting started or are revising a paper, it's good to think about the possibilities.

Checklist on how to organize and write a rough draft

1. Determine what questions you are answering for your audience and how specifically each question directs your approach.
2. List ideas that will convey the answers.
3. Arrange the ideas in an orderly sequence.
4. Using your own judgment, choose one of the four approaches (define, compare–contrast, enumerate, give cause–effect) and write immediately.
5. Recognize the need to revise.

References

Day, R. A. (1998). *How to Write and Publish a Scientific Paper*, 5th ed. Oryx Press, Phoenix, AZ.

Marshek, K. M. (1982). Transitional devices for the writer. In *A Guide for Writing Better Technical Papers* (C. Harkins and D. L. Plung, eds.), pp. 107–109. IEEE Press, New York.

Montgomery, S. L. (2003). *The Chicago Guide to Communicating Science*. University of Chicago Press, Chicago.

Tichy, H. J., and Fourdrinier, S. (1988). *Effective Writing for Engineers, Managers, Scientists*, 2nd ed. John Wiley & Sons, New York.

Zinsser, W. (1988). *Writing to Learn*. Harper & Row, New York.

4

SEARCHING AND REVIEWING SCIENTIFIC LITERATURE

"Do not condemn the opinion of another because it differs from your own. You both may be wrong."

DANDEMIS

PLANNING THE LITERATURE SEARCH

Any scientist needs to be familiar with the work that has been done in his or her area of research and that others are doing concurrently. No scientist, however, can read all the books, articles, and other printed matter related to a given area, perhaps not even all that are concerned with his or her research. Therefore, we need to be selective and to find the most pertinent literature on a subject as efficiently as possible, and use it effectively.

A deliberate, well-organized approach to a literature search can save you a great deal of time and energy. Searching the literature is a continual process, but first you will need to explore ideas related to your subject, conduct a specific search in relation to your research objectives, and certainly keep current with what is published while you are pursuing your objectives.

An **exploratory search** should be conducted before a clear hypothesis and clear objectives have been established. During this exploration, you will investigate the full scope of the subject area and determine topics of special interest to you.

Then, look for gaps in the research on those topics and decide what specific experimentation to propose and pursue. In your exploration of the literature and in the more specific search that follows, look for any duplication of your efforts. If someone has already done work similar to what you propose to do, you may need to alter your plans to avoid reinventing the wheel.

The **specific search** is essential for providing background to your own work and for revealing where and how your research fits into the scientific pool of information available. It will begin with a survey of information associated with your specific objectives. As your research progresses, you will encounter even more specific needs. For example, you may be using a particular method and need to know other uses that have been made of that method or what new equipment is available to employ with your procedures. In such situations, your search becomes limited to a single point.

Finally, when you are communicating your results, keep current with ongoing research. This **current search** can be maintained through habitual reading, interaction with your professional peers, and a periodic check of current awareness sources that can even e-mail to you titles of articles just being published.

In designing an overall plan for a search, consider the following principles.

Visit the Library

This admonition may sound facetious, but I'm quite serious. Your personal computer may be connected to a network whereby you can do most of your searching through your own office, but you will do well to visit the library first. Browse for 30 minutes, and check out the systems for finding and acquiring materials. Collect any flyers or brochures about the library and how it serves you. Determine where special materials such as government documents are filed and how to gain access to materials housed in other libraries. Interlibrary loans may be free, or a fee may be charged for acquisition of materials from other libraries. You may wish to browse through your library's collections or go online to browse at other libraries. A general familiarity with the library can be a real asset as you search the literature.

Allow Time

Often scientists are so interested in what is going on with the research itself that they neglect the literature search. You can actually save time in some instances by consulting the literature to discover the latest data compiled on a subject or a new technique for accomplishing your goals. More important than saving time is your responsibility for knowing what has been and is being done in your area. This knowledge is essential to scientific communication whether you are writing a proposal for new research, reporting results of a study in a poster, making a slide presentation, or writing the journal manuscript about completed research. Give yourself time to explore, find, and read about your subject and perhaps time to write a review of the literature.

Isolate Your Objectives

You can waste time when you spend too much of it exploring before you establish a specific search or if you repeatedly get sidetracked on interesting subjects that are related but not directly associated with your work. Once you have declared a hypothesis and specific objectives, stick to them in your search. At least be sure that any interesting sidelights are worthy of the time you give to them.

Document Carefully

Make detailed notes on important works that you find. For a future bibliography you may need authors' full names, full titles, journal titles, volume and page numbers, publishers, dates of publication, and any other information that is relevant to the style of the publisher to whom you submit a manuscript. See that any photocopies you collect have all this documentation on them. The ability to photocopy is a great convenience, but overuse of the practice can be detrimental to writing a review of the literature. Voluminous piles of copied materials can be just that—unorganized piles. Develop some system for filing photocopies so that important information does not get lost again after it has been found. Note cards are still handy for compiling information from the articles and for making the first draft of a bibliography. Or you may want to invest in a software program, such as EndNote or ProCite, or access a Web-based bibliographic management software, such as RefWorks, for compiling and filing your references. Your library may provide access to RefWorks free of charge.

Be Selective

Not everything written about your subject is relevant to your work. Some publications can be substandard, and reference to them can jeopardize your own credibility. Leaving out important works can be as bad as including questionable work. Discriminate relative to source, author, date, and relevance to your work. Other discriminating scientists are familiar with the literature and with the scientists who are working in a given area. They are not impressed by work padded with irrelevant literature or with that which fails to give reference and credit to the pertinent studies. Be especially careful in selecting literature from the Internet (Henderson, 2002). Your library has probably been somewhat discriminating in choosing books and journals that contain valid educational materials, and the databases they access via the Web are generally reliable, but the discrimination has to be entirely yours when you use the Internet independently.

Check for Accuracy

The most important principle that you can follow in searching and using the literature is **accuracy.** Careless use of ideas that allow ambiguous interpretation can injure your reputation. Careless documentation that leads fellow scientists to waste time pursuing a faulty reference should lead to your obliteration from the scientific community. Whatever style you use, be sure to include not only accurate but also enough information for a reader to find the source to which you refer. An accurate and complete reference is as essential to your writing as an accurate measure is to your laboratory research. To load your bibliography with what you think may be proper spellings of names or proper volume and page numbers is as dishonest as loading your scientific data. Verify all ideas and documentation.

Be Willing to Quit

Almost any library search can be extended indefinitely. Don't leave a search before the job is done, but don't continue to follow blind leads that give you fragments of remotely related material. At some point you must be willing to discontinue the search temporarily and apply what information you have gained to your research and to your writing. You can, and should, go back for more and for an update.

Create Something Useful

From your notes and documentation, put together a complete bibliography. Sort out articles and books directly related to your own study and write a draft of a literature review on the subject. This writing can be important to a proposal, a thesis, or a subsequent journal manuscript.

Verify

This principle is so important that it needs to be considered repeatedly as you collect the literature and when you use it. Check your bibliography and literature review against the original sources. Be sure there are no errors. Also, be certain that your list of references or literature cited at the end of your paper corresponds with the citations in your text after you have written it. Reviewers and editors can become unhappy with inaccuracies or inconsistencies between that list and what you have cited in the text. If you want to publish or be approved by a thesis committee, check these things carefully. Verify every time you revise or create new copy.

Keep Up to Date

We are all familiar with the college professor who is still lecturing from notes he or she made as a graduate student. Don't fall into this trap. Many thousands of books and articles are published every year on scientific subjects. Some of those relate to your work. Find them.

FINDING THE LITERATURE

In searching the literature, as with any job you do, you need to be familiar with the terminology and sources of information. Research findings and other scientific information can be published in books, monographs, proceedings, journals, dissertations, patents, standards, governmental bulletins or reports, and a variety of other sources. Primary source materials are published in various formats: paper, electronic, microfiche, video and audio tapes, and other media. Access to these primary source materials is available through indexes, databases, and catalogs with various searching techniques. Finding one source may require an online computer search; another, a paper index.

For a thorough search of the literature, you will probably do both manual and electronic searches. Before you do either, you may need to decide where to look—what databases to call upon and which indices to consult to find the specific literature you need. The electronic databases generally index and abstract more recent publications, and the paper indexes will cover previous years. Various guides such as the one by Schmidt *et al.* (2002) can provide information on appropriate indices, abstracting services, and avenues for finding the primary literature. Other guides, including those by Knisely (2002), McMillan (2001), and Smith (1998), provide insights into searching and using scientific literature. Your best guide is your reference librarian.

Most academic libraries use similar classification systems for shelving books, journals, and other materials, but each has its own unique characteristics. A journal you find at one library may not be in another. Libraries may provide access to databases from different vendors for specific areas. Most of them can now provide access to databases and catalogs through remote access. They have annotated listings of the databases they provide, explaining what information each can supply and the dates of coverage. Library science is a relatively complex discipline, but librarians with expertise in the field are usually eager to assist you with a search. Reference librarians can identify what is available for your search and how to access it, and they can recommend indices for manual and computer searching or thesauri for compiling keywords for your search. They can assist by showing you how to formulate search strategies, how to choose relevant keywords, and how to use Boolean operators. Find out what and who are available at your library and use these services. Many libraries will provide a specialist in your subject area. These specialists often teach classes and provide workshops or seminars on information literacy, sources of information in special disciplines, how to search specific databases, and instructions often tailored to coincide with class assignments. Take advantage of classes or special sessions on information literacy so that you can become proficient in searching the various resources and get maximum retrieval for your time and effort.

Computer databases are certainly a godsend for any literature search. Recognize all their advantages. They can usually be time efficient and comprehensive and generally access a wide range of the literature within a discipline. Acquiring the literature itself to read and incorporate into a review of your subject is also made easier with electronic access. Many journals are now published electronically; full-text access may be available to them through your library's catalog and through aggregated databases. New possibilities are becoming realities every day. Most libraries provide remote access to their holdings for their faculty, students, and staff and even provide access to other associated libraries. For electronic access to information and literature retrieval, be sure to keep up with what is available. However, what I say here may be outdated next year.

Although you can acquire voluminous information in minutes via computer, you should also recognize the limitations of database searches and compensate with a search outside their boundaries. Without proper planning, you may acquire citations that are mostly irrelevant to your specific interest, and it will take time to sift the grain from the chaff. In addition, most databases were not established until the late 1960s or the 1970s. The extent to which they have retrospectively converted information before these dates to a machine-readable format may be limited or nonexistent. Earlier works are still listed in printed abstracts and indexes. Consider the appropriate period to search relative to your subject. You want to be on the forefront of knowledge in your area, but that condition may require that you look in special publications not filed with an electronic database or find important work that was done before computer databases were established.

Consider an approach for your own search. Your specific search begins after you have explored a library and have established a hypothesis and specific objectives for your research. You or one of your advisors or colleagues probably knows of someone who has written about your subject in the recent past. A convenient way to start your search is to find that book or article that has been published recently and not only read it but also use any list of references it gives to find other articles. Use these other articles in the same way to create a **snowball** (Smith *et al.*, 1980). Each source yields another group of references, and in many instances, you can accumulate a great deal of material on a subject simply with a snowball search. Jot down authors' names and key words relative to your subject as you scan these references.

This approach is limited in that you are always going back in time, and the subject matter may get farther and farther from your own immediate question. This limitation may actually be an advantage when a manual snowball technique is used with searching databases that have not included information previous to the 1960s or 1970s. A second possible disadvantage to the snowball search is that sometimes scientists limit their references to persons closely associated with their own work or in a given society, and you may miss relevant publications in other countries, in other languages, or in sources outside the realm of those being cited by the group in your snowball.

After you have done a snowball search, you'll begin to feel comfortable with the terminology and names of the researchers who work in your area. Then you may do a database search to retrieve articles on your subject by using key authors, keywords, subject words you have identified to design a comprehensive database search. If you have had little or no experience with computer searches, a subject librarian can save you much time and frustration not only in the searching itself but also in planning for it. In searching the electronic databases, be sure to read the help screens; these will normally have excellent elementary and advanced search help.

When you have sorted through the relevant and irrelevant information from your electronic search, as well as the initial manual searches, you should have acquired a sizeable bibliography. However, if all your searching has not covered the full scope of time and content that you need, your next job is to hand search for materials outside those available electronically. Your initial searches may have missed isolated publications important to your research, such as work in progress but not yet published or information published in a book or pamphlet not covered by the databases or indices. You may be able to locate such references in a paper index or an online catalog or a database relating to research in progress such as CRIS. Current awareness sources such as Current Contents Search, Ingenta, or ArticleFirst will give you relatively recent information.

Several instructive guides are available for doing literature searches. For becoming familiar with information technology and the terminology used for online searches as well as elementary search processes, List (1998) and Munger and Campbell (2001) provide easy explanations and may be helpful in showing you effective search techniques. Along with providing a comprehensive guide to the secondary sources in the biological sciences, Schmidt *et al.* (2002) give information on online searches. It may appear to be outdated, but the guide by Smith *et al.* (1980) still provides the basic information needed for a manual literature search. Keep relevance and credibility in mind as you acquire information from any source.

SELECTING AND EVALUATING THE LITERATURE

Sources to Use

With the plethora of information available on any subject, it is vital that you select the sources that are relevant to your specific topic and reject the irrelevant or inappropriate. You will want to read completely all sources that are closely associated with your study. However, you will retrieve numerous references that are not so important to you, but you won't know that they are not until you scrutinize them. Develop a method for saving time and finding out whether you should read the entire article. If the title appears pertinent, read the abstract. If that still indicates information that may be helpful to you, read the conclusions and look at relevant data in any tables and figures or at a method that may be pertinent to your research. By the time you have done those things, you should know whether to read the entire article. Using this technique to screen articles can save you some time. But when you decide to cite a reference, you need to be thoroughly familiar with all that the authors are saying. Misquoting or taking a finding out of context can constitute an inexcusable inaccuracy.

Although it is often difficult to be selective, every reference you use should be credible and relevant to your own work. The literature may be relevant if it serves as a historical background to establish the position of your research in a larger framework or if it teaches you a new method or gives you new ideas to pursue in accomplishing your own objectives. It may illustrate or justify a specific point you make in your work, or it may support a result you find, a method you use, or a conclusion you reach. Include the citations that disagree with results from your own research. Determine why other researchers got different results and report the differences. An unbiased, comprehensive discussion of the literature will increase your own credibility.

Evaluating Sources

After you have collected all the literature you can find that is relevant to your study, you need to determine whether each source is an appropriate, credible reference for your literature review. How appropriate it is will depend upon how valuable the reference is relative to your audience and your objectives. Credibility or reliability of the source depends on the author, the publisher, and their purpose in publishing. Credibility is important no matter where you acquire a reference, but be especially cautious of information from the Internet. Consider the following points and the discussion of the numbered points that follows relative to each item you have collected no matter what the source is.

Judging Relevance and Credibility of Scientific Literature

1. Is the source useful for supporting or describing your objectives?
2. Is the date of publication timely and relevant to your topic?
3. What are the credentials of the author?
4. Who is the publisher? Was the document reviewed before publication?
5. Is the language unbiased and objective?
6. If it is a report or review about scientific research,
   a. is appropriate literature cited?
   b. are the methods scientifically sound?
   c. are the data objectively interpreted?
7. If it is an electronic source,
   a. who is responsible for the publication?
   b. does it have links to other credible sources?
   c. which domain is used for access?

---

1. The source is useful only if it fits your topic, but it does not have to agree with your hypothesis or with the results of your study. It is as important to show information about research that has found results contrary to yours as it is to show that which supports yours. What you want to eliminate is any literature that is just remotely related and not strictly relevant to your topic or any that is not reliable.

2. The date when the document was written or published is important. For the most part, you should include the most recent literature with the latest findings about a scientific issue. However, don't discredit work that was done years ago. For some subjects, studies from 50 or more years ago may reveal findings that are important to your topic. Timeliness of the literature depends on the subject itself and whether you are doing a historical review of the literature. For example, studies on transgenic plants were not likely done 50 years ago, but work on plant identification was and may be important in a study describing or naming species.

3. The author does not have to be a renowned scientist. Good work is done by junior scientists and graduate students. But check the affiliation of the authors. Are they associated with a university or a reputable agency that does unbiased work? Scientists associated with for-profit organizations or special interest groups can also be unbiased, but they likely favor the group that supports them.

4. Professional societies and reputable publishers usually publish credible reports. However, they may also publish news items or lay magazines that speculate on possible breakthroughs in science that have not been clearly established with sufficient data. You can generally judge these by the format of the document or the intended audience. It is important that scientific manuscripts be reviewed by other scientists. Scientific journals provide information on whether their publications are reviewed, and most of them are. If the publisher is not a professional society or a journal is not reviewed and well known, determine who that publisher is affiliated with and the reasons for publication. Language bias may give you answers.

5. Language bias is fairly easy to detect. If the author or publisher is trying to sell an idea or a product or is defensive or subjective about an issue, it is usually obvious in tone or the way things are written or not written. As Henderson (2002) says, "When in doubt, doubt." Be cautious with using any reference in which opposing views are not considered or any reference with emotionally laden language, with exaggerated or flowery language, with evident prejudice, and especially with no identified author or with conclusions that lack adequate supporting details and data.

6. Inaccuracy or incomplete disclosure can usually be detected in the scientific report simply by studying details in the literature, the methods, and the data. If authors fail to credit the appropriate literature or to include that which does not concur with their finding, their report may be accurate but incomplete or biased. Determine whether their methods are scientific, complete, and reproducible. Were there adequate samples and replications or repetitions in the study reported? Are the data objectively presented with applicable statistical design and analysis? Is the discussion objective and free of language bias?

7. The six numbered points above apply to any source of information whether it is from a scientific journal, book, an unpublished report or presentation, or any electronic document on the Internet. Be doubly cautious about information retrieved from electronic sources. A few additional clues are available to help determine credibility of electronic sources. First, determine what author and publisher are responsible for the information. Check any links to other sources. If the work has no obvious author, publisher, or links to other sources, question the credibility carefully. Check the domain in the access address or pathway, but don't depend on the domain for reliability. These domains are listed with abbreviations such as edu, com, gov, net, or org. Unreliable information can come from any of those sources. For instance, org may be the abbreviation used for the American Medical Association, but it may also be used for a subversive group using the Internet for propaganda purposes. In addition,

edu is used to designate both highly reputable scientists at universities and other employees of a university who may have an agenda of their own. The design of a Web page or other Internet document may also give a clue to credibility. Serious organizations or other publishers may present their logos or other attractive designs, but these are usually standard, subdued, and conservative. Graphic design and decorations, similar to language bias, can be used to call forth attention or emotion that can sell an idea or product. Be wary of such decorations.

Beyond these suggestions, Henderson (2002) and Munger and Campbell (2001) provide some good checklists for determining credibility of online as well as paper sources. All this close scrutiny and evaluation of sources should not take too much time. Not every detail has to be checked before you accept the information as reliable. Most reputable scientific journals are produced by professional societies or professional publishers, and they publish reviewed articles by reputable scientists. Scientific research done by government employees with the U.S. Department of Agriculture, the Department of Defense, or other agencies may be published as government documents. Generally you can accept these publications without much further evaluation, but distinguish the purpose and the intended audience for materials published even by these sources. These publishers may also print information in lay magazines that does not contain the complete data you need to support your own contentions. They may also publish quick communications, letters, or proceedings that are clearly designed to disseminate information before research is completed and ready for journal publication. Distinctions in primary publication of research findings and early reports on work in progress or speculations on interesting applications for the research are usually obvious. As a writer, you should be discreet in your use of these publications.

Some industries, companies, or corporations publish trade journals or information brochures that can contain valuable information, especially concerning what is available for doing research and development or instructions for use of a product that you need in your study. One must accept a bias in these publications toward the company, reflected by publishing the work without discrediting either that company or its competitors. Similarly, corporations may publish information for their employees to help them keep up with research and development. Such corporations or companies may also publish valuable educational materials for school children or the general public. Usually if you can determine the purpose for the publication and the intended audience, you can judge whether you should reference a source in a scientific paper. The scientific writer must use clear discretion in citing publications. Your own credibility is at stake.

Using Unpublished References

In addition to publications on work completed, know what is being researched on your subject. As you become involved with a profession, you will exchange information with your peers at professional meetings or by phone, electronic mail, or fax exchanges. You may sometimes use such personal communication as a reference in your own work. A bona fide, refereed publication is more likely to command the respect of your audience than is a personal communication. But verified communication from an expert on a subject can be valuable support to your paper. We can expect that the results from quality research should be and usually are published. If, however, the information is an isolated finding or a very recent disclosure, it may be important but not yet published. Personal communications can also supply you with information on papers in review or in press or with research projects underway. A fellow researcher can refer you to literature you have overlooked or can help you perfect a method you are using. In association with your peers, you soon learn who is respected, and you can aim for the same reputation. Your professional ethics must guide you in your openness as well as your confidentiality with your peers. Careful documentation of your sources is always vital.

THE LITERATURE REVIEW

Before you begin to write a review of literature, be sure you understand what constitutes plagiarism and how to avoid it (see Chapter 12). Document all information carefully, and follow a technical style sheet for specifications in textual citations and bibliographic entries (see Chapter 8).

Literature reviews bring together background information that integrates ideas from many sources or links other original research to your own studies. They can serve to justify your work, support your results, establish your methods, or simply add to knowledge on your subject. You will understand your own work better when you not only have read and studied what others have done but also have brought their ideas and yours together in writing. Literature reviews support proposals, theses, journal articles, oral paper presentations, and other reports. They may also be written to stand alone as review articles. Complete books, monographs, or journal articles can be made up of reviews of the literature. The suggestions that Day (1998) provides on writing review papers can be helpful in any literature review. See the sample review in Appendix 4.

Writing the review is no easy task. First, distinguish between the review that is published alone and the review of literature for your thesis, proposal, or journal manuscript. Just as with the literature search, your efforts should be based upon the goals for your writing and the objectives of your own research. Once you have searched the literature thoroughly, you read and become familiar with what you have found. Probably at this point you will begin organizing your notes. Think in terms of the organization that your review will take, and devise sections and headings for your notes or develop a complete outline to keep you on course.

For most graduate students, the literature review will become a chapter in your thesis or a review for a proposal or journal manuscript, and the content should be based upon your research objectives. Your notes might, for instance, fall into categories such as general information, methods in other studies, support for objective 1, support for objective 2, similar hypotheses, results to compare to mine, history of subject, pros and cons of controversy, and also may fit somewhere. Note cards or loose leaves that can be organized and reorganized are handy for these notes, but if you prefer a bound notebook or the computer screen, you can develop a system for keeping up with the arrangement of your information.

After you have become thoroughly familiar with the literature on your subject, the job of writing a review of it may look formidable. Complex jobs are usually done best if you divide them into parts and attack each part separately. I have two suggestions.

Develop an outline. Your subject matter will govern the kind of overall organization. If you are showing how your subject has developed over the years, choose a chronological arrangement. Comparison and contrast can be used in reviews of controversial theories on a subject. If your research brings together several views on a subject, you can enumerate and describe the various opinions. Or you can organize on the bases of the topics involved in your own work. Whatever your organizational plan, have it in mind before you write.

With a format in mind that will accommodate chronology or enumeration of points or other organizational design, look again at the headings you have given to your notes. You may now want to reorganize material, combine two parts, or separate one into two. In this reorganization, write down headings for the sections of your literature review. What you are doing, of course, is outlining. If the review is relatively long, you can organize the material for each section to make independent outlines of small portions of the review. You may want to make outlines of all the sections before you begin to write in order to see how they separate or connect. Parts can be combined or separated and reorganized yet again. All this organizing and reorganizing may seem time-consuming, but it may well save you some time in writing and rewriting and help to produce a better paper.

When you are satisfied with the organization of a single section, write that one, then another, and another until you have a draft of each. Then go back and put them all together being sure you have made comfortable transitions for the reader.

Create a skeleton. If outlines evoke screams of protest from you, do some freewriting first. Think about topics that are important to your study or review and write down ideas on these topics. When you feel that you are comfortable enough with putting those ideas down on paper, take a few of the documents you have collected and write a short paper based upon these. For a fair-sized review, such as one you might use in your thesis, as few as five or six articles will be enough for this initial effort. Keep in mind your own objectives, and organize the paper based upon the main points of concern in your study. Discard anything that is not going to serve your objectives and the final draft.

With this minipaper written, you can now isolate and refine the main points and set up the sections of your full review. Don't scream; you have to outline or at least organize sooner or later. Freewriting is for sorting things out so that they will start falling into place. Now the information you have freewritten from the few articles may need rewritten and arranged under the sections. This revision will serve as a skeleton of your first draft of the full review. You can now take other documents that you wish to review and add or insert information from them into what you have written about the first five or six articles.

This method works well with the word processor in that you are revising and editing as you add and rearrange ideas from your literature. Word processing also helps you sort out a stack of literature and get information from each piece into the right places in your text. You may wish to cite the same paper in more than one section of your review. With an initial skeleton of your work on the screen, you can add to one section and then to another as you glean information from each source. After you have inserted the ideas from various sources into your paper, you will certainly need to revise to smooth the flow of the language and the transitional progression from one point to another.

Outlining and creating a skeleton paper work quite well together. Whatever organizational technique you use, the draft will need to be revised to make the information flow to the reader in an organized fashion. You might find helpful the detailed steps Silyn-Roberts (2000) outlines to produce a literature review.

A common problem in writing the literature review is that it can turn into a boring list of ideas in paragraph form. Make yours move from one point to another with discussion and transition. Use subheads (not too many), transitional phrases, and unifying ideas to make information flow smoothly. Spice your writing with variety. Recall all the times you have almost fallen asleep trying to read boring material. Keep yours alive. Vary the way sentences and paragraphs begin. And be sure to use an assortment of active verbs. Literature reviews can quickly wear down a handy word. It is boring to read: "Author A found," "Author B found," "What Author C found was," "Author D also found." Other verbs can substitute for *found*. Try some of the following as well as others: "demonstrated," "presented evidence for," "suggested," "observed," "reported," "examined," "concluded," "noted," or "was convinced that" And vary the sentences by beginning with something other than the author's name: "In a study on," "Early in the 1980s," "Author A...," "According to Author A...." Our language is filled with active verbs and variations in sentence structure. Use them. Note the variety of sentence structures and active verbs in Appendix 4. Terry Gentry smoothly brings citations into the text, sometimes as a part of the sentence and sometimes less obtrusively in parentheses.

References

Day, R. A. (1988). *How to Write and Publish a Scientific Paper*, 5th ed. Oryx Press, Phoenix, AZ.

Henderson, J. (2002). "ICYouSee: T is for Thinking: A Guide to Critical Thinking about What You See on the Web." http://www.ithaca.edu/library/Training/hott.htm (verified July 11, 2003).

Knisley, K. (2002). *A Student Handbook for Writing in Biology*. W. H. Freeman & Co., Gordonsville, VA.

List, C. (1998). *Introduction to Information Research*. Kendall/Hunt, Dubuque, IA.

McMillan, V. E. (2001). *Writing Papers in the Biological Sciences*, 3rd ed. Bedford/St. Martin's, Boston.

Munger, D., and Campbell, S. (2001). *Researching Online*, 4th ed. Longman, New York.

Schmidt, D., Davis, E. B., and Jacobs, P. F. (2002). *Using the Biological Literature: A Practical Guide*, 3rd ed. Marcel Dekker, New York.

Silyn-Roberts, H. (2000). *Writing for Science and Engineering: Papers, Presentations, and Reports*. Butterworth-Heinemann, Oxford.

Smith, R. C., Reid, W. M., and Luchsinger, A. E. (1980). *Smith's Guide to the Literature of the Life Sciences*, 9th ed. Burgess, Minneapolis, MN.

Smith, R. V. (1998). *Graduate Research: A Guide for Students in the Sciences*, 3rd ed. University of Washington Press, Seattle.

5

THE PROPOSAL

"A new idea is delicate. It can be killed by a sneer or a yawn;
it can be stabbed to death by a quip and worried to death
by a frown on the right man's brow."

CHARLES D. BROWER

Your ability to write a research proposal can be vital to obtaining a degree, getting
a job, and advancing to higher positions. As a scientist, you will encounter two
important kinds of proposals: the graduate research proposal and the grant proposal. Distinctions in the two exist because of different audiences, guidelines,
and purposes, but both require similar planning, composition, and execution.
The graduate proposal is a plan for you and your committee to follow relative
to your research objectives. Your basic purpose in writing the grant proposal
is to obtain funds to pursue certain research objectives. Developing skills in
writing proposals is especially important for the young scientist who has not
yet established a professional reputation.

THE GRADUATE PROPOSAL

A written graduate proposal serves at least three purposes. First, it is your
communication with an advisor and a committee. You are asking for approval
and support to pursue a project. When the answer is yes, then the proposal
serves as an agreement between you and those granting the approval. The third
purpose for a proposal is especially important to you. The contents become

your plan of action and will serve as an outline of work to be done throughout the project.

Whether or not a written proposal, or prospectus, is required for a graduate research program, one should be written. For a graduate student, important reasons for organizing and writing your proposal are that the writing helps you to plan your work in advance, to review what has already been done in the area, to foresee the pitfalls that lie ahead of you, and to remain focused on a reasonably direct track between your proposed objectives and your goals. A very practical side effect in writing a graduate research proposal is that some of it will serve as drafts for your thesis and for journal articles. With modifications, the introduction, review of literature, and methods will become parts of your thesis. Without a clear plan or written proposal, the time needed to acquire a degree can be increased by a semester, a year, or even more.

But a graduate student proposal is not just a plan for you to follow; it is also a commitment to a departmental research program, to the graduate school that admitted you, and to your advisor. Your proposal is your commitment to make wise use of the time and resources available to you. Often in the scientific disciplines, the advisor, as well as the department and graduate school, has made an agreement with you to direct and financially support your research. If you fail to live up to your proposed commitment, you cost others time, effort, money, and progress in a research program. Your proposal gives those with interests in your accomplishments some assurance of success by presenting a goal to be accomplished, rationale and justification for pursuing that goal, and feasible methods to accomplish the goal. Your job then is to accomplish that goal. As DeBakey (1978) says, "It is a mistake to promise mountains and deliver molehills."

The Grant Proposal

The same admonitions that are true for the graduate proposal are also true for the proposal to be sent to a granting agency. The audience is different and purposes may be motivated more by the stage of your career and by the desires of the granting agency than by the need for research education and the obtaining of a degree. But similar content and form and the same principles for writing are required. I suggest that the writer of the grant proposal carefully explore the wishes of the granting agency, be sure to follow its guidelines, and refer to a source such as Reif-Lehrer (1995) for expert recommendations for grant writing. Her experience is largely with big granting agencies such as the National Institute of Health and the National Science Foundation, but if you can follow suggestions for success with proposals to these groups, you can also write for other grantors. I understand that about the time this book is to be published, Reif-Lehrer will have a new edition of her book available; you may want to check on that.

Before you take pen in hand or put fingertips to keyboard, be sure you are ready to write a grant proposal. It will take you far less time to write a proposal if you prepare for it thoroughly. The likelihood of success will increase with this preparation and with allowing adequate time for both preparation and writing. Be sure your idea is good and is fitting for what the funding agency wants. Discuss it with colleagues and perhaps with the funding agency. Your reviewers will probably be scientists; your proposal must be scientifically sound. Study the topic. Outline a plan and review it carefully. Consider what personnel, money, equipment, and time it will take, and consider how this research will fit into the rest of your work load. Check with your own institutions to see what their requirements are and know how to fulfill them. Plan to have colleagues review the proposal; set them up in advance with a tentative date you will have it to them—well in advance of your deadline for submission. Review the granting agency's guidelines; study Reif-Lehrer's (1995) checklists; talk with colleagues who have had proposals funded. Devise a schedule; set deadlines for studying the literature, completing a first draft, getting it to reviewers, revising, and meeting the submission deadline. Then write the proposal.

Content and Form

Before you begin to write any kind of proposal, you must have a good idea and clarify your thinking, planning, reading of background literature, and formulating of hypotheses and objectives to have appropriate content to put into a format. DeBakey (1978) rightfully suggests that you write down the precise question you are setting out to answer and the expected results. Solidify your hypothesis and objectives, and clarify what you want this research to achieve; those issues will guide you through the other parts of the proposal.

Whether your proposal is a graduate research proposal or a grant proposal, form and content will depend on your own intent and on the audience to which the plan is directed. Private and corporate foundations and government agencies that support research often supply detailed guidelines specifying the order and length of the sections in a proposal; information that must be included; and even specifications for the type, type spacing, and other elements of style. It is essential that you follow such guidelines. Many grants are competitive, and a proposal can be rejected simply because the author does not follow instructions. When a grantor receives 200 proposals and can provide grant funds for 10 of those, every detail is vital, including the appearance of the document. Take a look at other proposals that have been approved in your area of research and perhaps establish a pattern based on them. Also, determine how the grant proposal should be submitted; some are now sent electronically. Be sure you have the latest update of guidelines from the grantor. The guidelines used in an earlier proposal may be outdated.

Your graduate school, department, advisor, or granting agency may suggest a format or outline, but often you will have to devise your own organization. You can write a successful proposal by considering the characteristics common to all and by setting up an organization that will best convey those characteristics to your particular audience. Remember that communication is essentially a question/answer process. Producing a successful proposal requires knowing what questions your audience will ask and answering them effectively. In other words, first consider your audience and the following criteria that will be used to judge your proposal:

1. **Originality and scientific merit or benefit to the grantor**
2. **Importance to the discipline or the immediate problem**
3. **Feasibility**
4. **Rationale and methodology**
5. **Ability and experience of the investigators**
6. **Budget, facilities, and time required**
7. **Appearance and adherence to guidelines**

Your proposal should answer several questions about a specific subject. Is it worthwhile? What are the chances of success? Are the investigators qualified to do the work? What benefits will be derived? Are the expenditures of time and money realistic? The answers to such questions constitute the rationale or justification that serves as the basis for both the proposal and the subsequent research.

Initially, as a graduate student, be sure the topic, your research question or hypothesis, and the objectives are absolutely clear to you and that your advisor approves. Once you have identified the questions you are expected to answer or the objectives you plan to pursue, you are ready to decide what sections to set up to accommodate those answers. Almost any proposal will include at least the first six of the following conventional parts:

1. **Title page and executive summary or abstract**
2. **Purpose or hypothesis and specific objectives**
3. **Discussion of significance or need (justification)**
4. **Review of work done or being done (literature)**
5. **Materials and methods**
6. **Discussion of possible outcomes (conclusions)**
7. **Time frame, budget, and biography of investigator(s)**

This list is **not** an organizational outline but a mere listing of what you should include. With these parts in mind, you can begin to set up an outline for yourself or follow the format imposed by your committee or granting agency. The best organization for one proposal may not be the best for another. You may begin with a discussion of the need for the research rather than with your purpose or hypothesis; you can even discuss possible outcomes in the introduction. You could start with a list of objectives and then build a case for pursuing them.

A proposal for a graduate student's research may require a far more comprehensive literature review than the grant proposals would permit. However, you may not need the budget, time frame, and biography. You and your advisor or committee will have considered those questions relative to your program. But you should become familiar with the format that requires these inclusions because they can be crucial for a proposal written to acquire grant funds in your future. Regardless of what format you use, you should include the following parts in whatever sequence concurs with your guidelines.

Title and Title Page

The title should identify the specific subject in as few words as possible. It should attract attention to the hypothesis and clearly reflect the objectives of the proposal. Write a working title before you write the proposal to give yourself a succinct, clear focus; then write the executive summary. When you have completed the proposal, scrutinize the title and executive summary carefully and rewrite them as needed. They are the crucial first impression you make on reviewers of your proposal. In the title, use only key words and avoid generalities or abstractions such as "A proposed study of the...." We already know it is a proposal.

The title page is also an important first impression. Be sure it is neat. In addition to a carefully worded title, this page will name the principal investigators (authors) and give their addresses, the date of submission, and the committee or agency to which the document is submitted. Many formats for grant proposals also require that the title page include information on the amount of funding requested and the time frame in which the work will be done. They sometimes require official signatures from not only you but also a company or institutional official who coordinates funding activities. Some groups, such as governmental agencies, provide a rather complex form to fill out as a cover page for the proposal. The cover form in Fig. 5-1 for the National Science Foundation is followed by a certification page that requires signatures of all principal and coprincipal investigators and legal statements or questions to answer regarding such things as contracts, cooperative agreements, debt, and disbarment, all of which you must sign. If you have no guidelines, or especially for the graduate research proposal, create your own neat title page (Fig. 5-2).

Executive Summary or Abstract

For the grant proposal, the executive summary or abstract is the most important impression you make beyond the title page. Make it concise but compelling. Its

COVER SHEET FOR PROPOSAL TO THE NATIONAL SCIENCE FOUNDATION

ROGRAM ANNOUNCEMENT/SOLICITATION NO./CLOSING DATE/If not in response to a program announcement/solicitation enter NSF 99-2	**FOR NSF USE ONLY**
	NSF PROPOSAL NUMBER

OR CONSIDERATION BY NSF ORGANIZATIONAL UNIT(S) (Indicate the most specific unit known, i.e., program, division, etc.)

DATE RECEIVED	NUMBER OF COPIES	DIVISION ASSIGNED	FUND CODE	DUNS # (Data Universal Numbering System)	FILE LOCATION

EMPLOYER IDENTIFICATION NUMBER (EIN) OR TAXPAYER IDENTIFICATION NUMBER (TIN)	SHOW PREVIOUS AWARD NO. IF THIS IS ☐ A RENEWAL ☐ AN ACCOMPLISHMENT-BASED RENEWAL	IS THIS PROPOSAL BEING SUBMITTED TO ANOTHER FEDER AGENCY? YES ☐ NO ☐ IF YES, LIST ACRONYM(S)

NAME OF ORGANIZATION TO WHICH AWARD SHOULD BE MADE	ADDRESS OF AWARDEE ORGANIZATION, INCLUDING 9 DIGIT ZIP CODE
AWARDEE ORGANIZATION CODE (IF KNOWN)	

NAME OF PERFORMING ORGANIZATION, IF DIFFERENT FROM ABOVE	ADDRESS OF PERFORMING ORGANIZATION, IF DIFFERENT, INCLUDING 9 DIGIT ZIP COD
PERFORMING ORGANIZATION CODE (IF KNOWN)	

IS AWARDEE ORGANIZATION (Check All That Apply)
(See GPG II.D.1 For Definitions)   ☐ FOR-PROFIT ORGANIZATION   ☐ SMALL BUSINESS   ☐ MINORITY BUSINESS   ☐ WOMAN-OWNED BUSINES

TITLE OF PROPOSED PROJECT

REQUESTED AMOUNT $	PROPOSED DURATION (1-60 MONTHS) months	REQUESTED STARTING DATE	SHOW RELATED PREPROPOSAL IF APPLICABLE

CHECK APPROPRIATE BOX(ES) IF THIS PROPOSAL INCLUDES ANY OF THE ITEMS LISTED BELOW

☐ BEGINNING INVESTIGATOR (GPG I.A.3)
☐ DISCLOSURE OF LOBBYING ACTIVITIES (GPG I.D.1)
☐ PROPRIETARY & PRIVILEGED INFORMATION (GPG I.B, II.D.7)
☐ NATIONAL ENVIRONMENTAL POLICY ACT (GPG II.D.10)
☐ HISTORIC PLACES (GPG II.D.10)
☐ SMALL GRANT FOR EXPLOR. RESEARCH (SGER) (GPG II.D.12)
☐ GROUP PROPOSAL (GPG II.D.12)

☐ VERTEBRATE ANIMALS (GPG II.D.12) IACUC App. Date _____
☐ HUMAN SUBJECTS (GPG II.D.12)
  Exemption Subsection __ or IRB App. Date _____
☐ INTERNATIONAL COOPERATIVE ACTIVITIES: COUNTRY/COUNTRIES _____
☐ FACILITATION FOR SCIENTISTS/ENGINEERS WITH DISABILITIES (GPG V.G.
☐ RESEARCH OPPORTUNITY AWARD (GPG V.H)

PI/PD DEPARTMENT	PI/PD POSTAL ADDRESS
PI/PD FAX NUMBER	

NAMES (TYPED)	High Degree	Yr of Degree	Telephone Number	Electronic Mail Address
PI/PD NAME				
CO-PI/PD				
CO-PI/PD				
CO-PI/PD				
CO-PI/PD				

NSF Form 1207 (10/98)   Page 1 of 2

FIGURE 5-1

Sample cover sheet for proposal to a federal agency.

content is the basis for a decision to consider or reject the entire proposal. The overall neatness of the proposal, the extent to which it follows a prescribed format, and the content of this summary will probably determine whether yours makes the first cut in a group of competitive proposals. For those 200 competitive proposals submitted, the executive summary may be the only section read before the 200 are reduced to a lesser number to consider for funding.

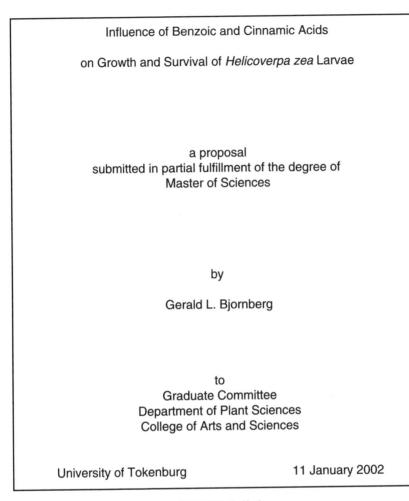

Influence of Benzoic and Cinnamic Acids

on Growth and Survival of *Helicoverpa zea* Larvae

a proposal
submitted in partial fulfillment of the degree of
Master of Sciences

by

Gerald L. Bjornberg

to
Graduate Committee
Department of Plant Sciences
College of Arts and Sciences

University of Tokenburg   11 January 2002

FIGURE 5-2
Typical cover sheet for graduate research proposal.

Your graduate committee may not require this summary, or the granting agency or your graduate committee may request an abstract rather than an executive summary. Used with the proposal, these two forms may be synonymous, but with either term what may be expected is simply a concise description or summary of the proposed hypothesis, objectives, and expected outcomes. For the graduate proposal, you should be able to find a successful proposal written by a former student and pattern the abstract or summary after the example if one is needed. For those of you seeking grant funding, try to acquire a copy of a successful proposal that has been submitted to your granting agency, or call personnel authorized to handle the proposals for the agency to ask for information. If you cannot obtain specifications, fulfill the request for an abstract or executive summary by following the characteristics for the informative abstract (Chapter 10) to whatever extent you can. The difference in this summary for a proposal and an informative abstract written for journal publication is that complete results cannot be included. This omission allows more space for justifications and methods. Begin with a sentence or two of justification followed by the objectives, a concise statement of the methods, and then the conclusions to reiterate justifications and emphasize benefits. Be sure your summary or abstract is well worded and establishes the credibility of the proposal and the investigators.

Introduction

The organization of the introduction will differ depending on the audience and the development of the full proposal. Whatever form it takes, it should immediately show the reader the subject to be investigated and give a rationale for pursuing the research. By way of identifying the relative scientific merit of the research and justifying its pursuit, the introduction may include some literature review and statement of benefits. It should include a purpose and suggest the scope of the proposed research. Above all, it should define the hypothesis and list the objectives.

Don't let the word **hypothesis** disturb you. As used here, the word simply refers to the proposition, the purpose of the study, the assumption you expect to prove, the question to be answered with the research, or the problem to be solved. It points toward what ideas can be credited or discredited when the objectives are satisfied. The **objectives** are specific goals. They should encompass the aims of the research, yet be brief, precise, and limited in number and scope. Trying to include too many primary and secondary objectives can obscure the focus of your proposal for you and for those who review it.

Definitions for hypothesis and objective are not always used with the same specific meanings that I have attached to them here. Although many, including Stock (1985), define the terms as I do, Friedland and Folt (2000) pretty well reverse those definitions. Others may use terms such as goals, aims, questions, and purpose rather than hypothesis and objectives. And some will require that you pursue a null hypothesis rather than a positive hypothesis. For example, if you hypothesize that a certain gene controls skin color, your null hypothesis would be that that gene does not control skin color, and you would try to prove that it does not. If you could not prove it, then a positive hypothesis would be true. Some researchers feel that trying to prove a null hypothesis keeps them more objective and less prone to overlook or discredit findings that might prove the positive hypothesis wrong. But all these differences in definition and in words used to establish a purpose in a proposal need not leave you frustrated. Simply check guidelines for your own proposal, ask your advisor for an opinion on what terms to use, refer to proposals by previous researcher that have been approved, and then stick to a consistent terminology in your own.

For clear justification and content in the proposal, be sure to distinguish between the hypothesis and the objectives in your own mind. Let me illustrate my definitions by following van Kammen (1987) in using Christopher Columbus as an example. If Columbus were writing a proposal for Queen Isabella and King Ferdinand, his title might be "Mapping an Alternate Trade Route to the Orient." His hypothesis would be that the world is round and that by sailing west he could reach the East and establish a new trade route for Spain and the rest of Europe. His objectives might be (1) to sail west and chart a route to compare to other routes and (2) to bring home three shiploads of spices. Notice the differences in the hypothesis and the objectives. The hypothesis is the general supposition and contains a preconception not yet proven. The objectives will determine whether the hypothesis is true, and they are the specific goals that will be achievable, if the hypothesis is true, by acquiring empirical data (distance and direction measurements) and tangible amounts of tea, cinnamon, and frankincense. These objectives have to be realistic and appeal to the grantor. Queen Isabella may have questioned how realistic they were, but both objectives would certainly be appealing to her. The same principles are true in developing a research proposal that appeals to an advisor and graduate committee or a funding agency.

Incidentally, while we are using Columbus as an example, let me call your attention to the observation that van Kammen (1987) makes about the "Columbus paradox." Christopher did not accomplish his objectives; his research hit a roadblock. He did not prove his hypothesis (although he thought he did), but his work made a definite impact on the Americas and the world. Further research on the same hypothesis did reveal a positive conclusion. You have not failed at all if your research leads you to the West Indies rather than to China. Your discovery may be more important than if you had accomplished your original intent.

Justification

Justification, or rationale, is the key word around which a proposal is built. It is the basic criterion by which the final proposal is judged. Justification permeates the entire proposal from the title to the conclusions with an appeal for approval and with evidence that the proposition should be pursued. Whether or not a specific section is given this headnote, be sure that all sections contribute to the rationale that justifies the time, effort, money, and other support necessary to accomplish your objectives.

Justification outlines what can and should be done to accomplish beneficial outcomes. It shows how the methods can accommodate the objectives and how satisfying each objective will help to achieve the final goal. Justification describes the importance of your research to science and to the application of science. You can also justify your project in terms of its timeliness and economic significance as well as your own ability and access to resources to accomplish meaningful objectives. Whether it is a section unto itself or an integral part of all sections in the proposal format, the justification will be based upon the following:

1. **Reason and logic**
2. **Preliminary research**
3. **Scientific principles**
4. **Previous research (literature)**
5. **Feasibility of methods**
6. **Use of or benefit from the results**

In some instances, your research may be partially justified on the basis of preliminary research you have done on a subject. If so, you may wish to present data you have collected, results that have come from your research, and a discussion of how and why these results indicate that further study has merit. In fact, some academic departments do not consider your proposal until you have pursued some preliminary research, and some granting agencies are more likely to approve your proposal if you present results of a background investigation in your proposal. These research results still need to be supported by scientific principles and the literature. The example in Appendix 5 represents this kind of graduate proposal.

Literature Review

The literature review should consist of a summary of ideas pertinent to your research. It can review the history of your subject or the present state of the art; it ought to consider any controversies surrounding a question or gaps in available information that you intend to fill with your research; it may introduce methods that will make your work possible. Whatever kind of support it gives your proposal, a good literature review can establish your credibility and your chances for having a proposal accepted by illustrating that you know what has been done, what is being done, and what needs to be done in an area. You should cite work being done and recent publications from other institutions as well as from your own to establish the relationship between your proposed research and that of others. Be sure that all discussion is relevant to the specific objectives and the general hypothesis. For most proposals, keep all sections, including the literature review, brief to hold the reader's attention to your objectives. Some advisors expect a lengthy review of the literature in a graduate proposal because they want you to study your subject extensively to understand your own research better. The literature review may not be a section in itself but a part of the introduction and included as citations appropriately placed in methods or discussion and conclusions. Ask your advisor or follow guidelines and examples of previously approved proposals.

Methods

In the review of literature or in the methods section, you will increase your credibility if you point out methods that you or other researchers have used with or without success. The methods section, often called "plan of operation," "materials and methods," or "experimental procedures," is the very foundation of the scientific merit and feasibility of the work. To convince your audience that your plan is feasible and to serve you in pursuing the research, the methods section should outline the working plans in as much detail as possible. Include information on materials, sampling, analysis, data, and even people you will need to work with; steps you will take in conducting the research; data you will collect; and how you will analyze and use the data collected. Describe any limitations or potential problems you may encounter and tell how you will address them. Procedures should generally follow the same order as the objectives and show how each objective will be attained. All materials and time needed should be justified in this section and reflected in the budget of a grant proposal. To the granting agency, this section should justify the budget.

Conclusions

Although you may have no results and little discussion beyond that described in previous sections, it is important to reemphasize objectives, summarize points in the justification, and draw the reader back to the research question, the hypothesis, and the objectives. Your conclusions will summarize points of justification and benefits to be derived. To further justify your proposal, this section can extend into proposed applications or future research beyond your own, but don't overdo this idea.

References

The reference section is essential to your proposal. References with full titles indicate the extent to which you have explored your subject and are helpful to reviewers in their considerations. Although all important sources should be listed in your references and cited in the text, padding with citations only incidentally

concerned with your proposal will not impress your reviewers. Both your graduate committee and reviewers for grant proposals are likely familiar with the
literature published. In fact, some reviewers may be authors of papers you have
cited, and they will quickly recognize omissions, padding, or errors in your references. The citations and references must be accurate and follow a consistent
style throughout. Any error can destroy your credibility and diminish the chance
for acceptance of your proposal. You may want to explore Chapters 4 and 8 for
further suggestions about the use of literature.

Budget and Time Frame

For a grant proposal, the most important issues can be time and funding
requests. A grantor may be fully convinced that your scientific proposition is
feasible, but an unrealistic budget or time frame can negatively influence opinions of reviewers. Just as some people ask for too much time and money, some
do not ask for enough. Be realistic in your needs. As DeBakey (1978) says,
"honesty and common sense are the best guides."

Plan the budget carefully with estimates that are as nearly accurate as possible.
You don't know exactly what your supplies will cost, but if your research has been
clearly outlined, you can provide a reasonable estimate. Include salaries, equipment and supplies, publication costs, travel to collect samples or to take results
to professional meetings, and even phone bills. Don't forget indirect costs or
overhead if your company or institution charges such a fee. Many times overhead
is a large percentage of the total costs; 20%, 40%, and even higher percentages
for overhead are not unusual. On the other hand, some agencies may not allow
such indirect costs to be budgeted. Be sure your budget can reflect an agreement
between your employer and the funding agency.

Some grants are made only if the employer matches the value of grant funds
in money or other resources. In this situation, you will need to include in your
proposal a statement on these contributions. Describe the facilities and equipment that your employer will contribute to carry out the project. Also note the
percentage of time the investigators and other staff or hourly workers will
spend with the project and consider percentages of these salaries as part of the
employer's contribution.

Time may be as important as money to your grantor. Allow yourself enough
time, but don't be wasteful. In asking for time, justification in the body of your
proposal is critical. Some experimentations takes longer than others. If you are
doing laboratory analysis that takes just 5 weeks for each data set and you are
supplying 10 data sets, you might reasonably complete your proposed research
and all the reports involved in 1.5 to 2 years. However, if you are doing field work
that must be repeated over the environmental conditions of 2 or 3 years, you can
justify additional time. With both time and money, the key word is realistic.

As a graduate student, you may not need to worry about your budget and time frame. Your advisor does enough worrying for both of you. You simply accept the support available, and your proposal concentrates on what you will do rather than how it will be financed. The time frame may be dictated to you. At some schools, you must finish your program within a certain time or your research support will no longer be available to you. Consequently, you work within the limits provided for time, money, and graduation dates.

Biographical Information

For your graduate proposal, you will not need to provide biographical information because your advisor has already looked into your credentials with your application to graduate school. But for the grant proposal, what is often known as a **curriculum vitae** is usually essential. Be sure to make clear in this abbreviated resume that you are capable of doing the proposed research. Emphasize points in your training and job experience that have to do with the expertise needed, and avoid an extensive presentation of points unimportant to the research.

OTHER CONSIDERATIONS

Proposals are likely to become a vital component in your career as a scientist. Be conscious of the requirements for a good proposal as you write your graduate prospectus. This practice will serve you well when you begin to write grant proposals to support your own research. In graduate school, if you are a paid graduate assistant or a research assistant, you may assist in writing grant proposals other than that for your own research. For example, you might be asked to assist with a proposal extending research beyond the limits of your own program or pursuing another avenue of research in the same discipline. Producing any proposal can be a valuable experience for you, and it will teach you much about the proposals you will produce in the future.

As with any other important document you compose, your proposal should go through a process of reviews and revisions. Often you must meet deadlines for submission of proposals. Write drafts in plenty of time for reviewers to make suggestions and for you to revise the proposal.

Whether a proposal is invited or competitive, be neat and direct, follow guidelines, and be realistic. Your ability, the time needed, the resources available, and the money granted must all add up to probable success from a proposed project. If these things are not treated realistically, all the good scientific ideas in the world will not yield as many positive side effects or as much funding for your research.

Despite all the work you put into writing proposals, many of them will be rejected. Reduce your frustrations by recognizing the beneficial side effects to writing unfunded grant proposals or even the graduate research proposal that your committee does not approve. Something positive often comes despite the rejection. Think of it this way. You have a good idea for scientific research. In writing the proposal, you pursue the subject in depth. You learn much from the literature, you learn who else is working in the area of research and often get acquainted with people you will work with in the future, and you practice your skills in writing proposals. In the face of a $150,000 rejection, these benefits may seem trivial, but in the whole picture they are important to your career. Proposal writing is a way to keep up with what is going on; it prompts you to read the literature and to consult and cooperate with your colleagues. In addition, the rejected proposal should serve as a semifinal draft for resubmission to your graduate committee or to the same funding agency or to another. A resubmission may be more likely to be funded than the original one (Reif-Lehrer, 1995). Often the funding agency, and certainly the advisor or graduate committee, will provide review comments that can help you in revising the document for submitting a more acceptable version on the second or third time. Consider revision and resubmission a part of the overall grant writing process.

Finally, as important as obtaining the approval or funding for your research is the execution of the project. As you get into the work, you may find that you have to revise the proposed plan. For instance, you will find a different method or new equipment that provides better results, you will discover evidence that must alter your original hypothesis, or specific equipment will become inoperable just when you need it. Your original proposal will not guide you through all the possible problems and discoveries that occur. You must be creative enough to find a new path and execute the research regardless of obstacles. Explanation of these obstacles and how you overcome them may have to be incorporated into progress reports.

PROGRESS REPORTS

To keep their records updated and because the unpredictable is to be expected, many grantors and graduate advisors require periodic progress reports. These reports protect your credibility. Be sure you get the reports in on time and clearly explain the state of the project, what has been accomplished, and the proposed next steps. If you have had problems, report them and do not be too pessimistic. Be positive, but as with all the details in your proposal, be realistic. Instead of saying, "We were not able to complete this experiment by April," say, "We should complete this experiment by June." And then explain any hurdles that prevented your finishing by April and how you plan to get around them. Grantors and advisors are generally reasonable people. They need to know that you are doing the best job possible in carrying out your proposed project.

The progress report may be as formal and as important for a granting agency as the proposal. It may be as difficult to write. Granting agencies may want quality assurance that research is being carried out with a sound scientific accuracy and that their funds are being used appropriately. They may renew your funding for an extension of the project if they like what they see. Again, be positive. If you have faced obstacles and overcome them, they may admire your efforts; if you've given up in the face of obstacles, they may not want to risk their funds with you again. Silyn-Roberts (2000) provides some good suggestions for preparing progress reports. The intermediate progress report will also be valuable as you write the final report.

A final progress report can also help you with that final step in any research project—the publication or thesis or the new grant proposal. No research project is complete until this final report is done. This report is evidence of your accountability or evidence that the time, money, and resources have been used well. Among the characteristics that are scrutinized in a beginning scientist is the potential for good grantsmanship or "the art of getting financial support (a grant) for your research" (Stock, 1985). Your ability to communicate through the proposal and in a final thesis, report, or publication and a new proposal is an important component in this art and absolutely vital to your career as a scientist.

If you are not following a prescribed format, you should explore differing opinions of others before you begin. Check what Peters (1997) has to say about graduate proposals. Silyn-Roberts (2000) offers good information on writing proposals and progress reports. Reif-Lehrer gives detailed suggestions for writing grant proposals to large funding agencies. DeBakey (1978) and O'Connor (1991) discuss basic strategies for proposal writing, and Meador's book (1985) contains basic principles. A sample graduate proposal is in Appendix 5. Regardless of whom you consult for information, be sure you follow the guidelines set forth by your graduate committee or the funding agency. Those guidelines may differ from the suggestions I or any other reference may make.

References

DeBakey, L. (1978). The Persuasive Proposal. In *Directions in Technical Writing and Communication* (J. R. Gould, ed.), pp. 25–45. Baywood, Farmingdale, NY.

Friedland, A. J., and Folt, C. L. (2000). *Writing Successful Science Proposals.* Yale University Press, New Haven, CT.

Meador, R. (1985). *Guidelines for Preparing Proposals.* Lewis, Chelsea, MI.

O'Connor, M. (1991). *Writing Successfully in Science.* HarperCollins Academic, London.

Peters, R. L. (1997). *Getting What You Came For: The Smart Student's Guide to Earning a Master's or a Ph.D.* Farrar, Straus & Giroux, New York.

Reif-Lehrer, L. (1995). *Grant Application Writer's Handbook.* Jones and Bartlett, Sudbury, MA.

Silyn-Roberts, H. (2000). *Writing for Science and Engineering: Papers, Presentations and Reports.* Butterworth-Heinemann, Oxford.

Stock, M. (1985). *A Practical Guide to Graduate Research.* McGraw-Hill, New York.

van Kammen, D. P. (1987). Columbus, grantsmanship, and clinical research. *Biol. Psych.* **22,** 1301–1303.

6

THE GRADUATE THESIS

"Camante, no hay camino, / Se hace camino al andar."
(Traveler, there is no path, / Paths are made by walking.)

ANTONIO MACHADO

THE THESIS AND YOUR GRADUATE PROGRAM

Except for works of rare genius, a thesis or dissertation (I use *thesis* to mean either) cannot be produced in a week or even a month. No designated time can be set for work on the thesis—some take 6 months, others a year. Preliminary work can add another 2 years. To make the most conservative use of time, the thesis should be written as the graduate program progresses. This time must be coordinated with activities such as research, course work, and professional meetings. Early in your graduate program, develop a time line with dates beginning at your entrance in graduate school and ending with your completing the degree. Along that line will be as much detail as you can enter, and you will revise it as you proceed through your degree program. Figure A6-1 in Appendix 6 may be helpful in designing your thesis time line, but in addition to the thesis and related work, add times for course work, details on steps in the research itself, graduate exams, job search, presentations at meetings or in your department, and whatever else fits your individual program. If you find it hard to juggle all these activities, read Molly Stock's (1985) book, *A Practical Guide to Graduate Research.*

The thesis is the document that records the research efforts in your graduate program and the results of them. Characteristically, your thesis should be built on:

1. **A complete library search**—on everything that has been done on the specific subject and closely related subjects
2. **Your original research or professional project**—field and laboratory experiments based upon a research proposal or project as approved by your major professor and graduate committee
3. **Your syntheses**—putting together and deriving meaning from data, ideas from others, and your own conclusions

To neglect any one of these foundations for your thesis can limit the quality of your work. Success depends on your knowledge of what others have concluded from their studies, meticulous work with your own research, and your efforts to give a clear, accurate perspective to the entire study. When these points have been given appropriate attention, the other criterion for success is communication—the writing itself.

The master's and doctoral theses differ, and expectations from one graduate school to another are inconsistent, even in the same discipline. Specific requirements are set forth by the department, the college, and the graduate school from which you get your degree. As you begin your graduate project, check on any criteria for theses from your graduate school and department, talk with your advisor, and peruse several theses that have been produced in recent years by reputable degree candidates in your department. In other words, get a clear feel for what your thesis should be long before you write it. Typically, the thesis will include the following:

1. *Introduction*—general justification for the study, the hypothesis or purpose behind the study, and a specific statement of objectives

2. *Literature review*—a detailed report from your library search about what has already been done on your subject (sometimes combined with the introduction)

3. *Materials and methods*—an account of the specific techniques used in the study, including materials needed, statistical designs, and data collection and analyses.

4. *Results*—a presentation of the data acquired from your research

5. *Discussion*—significance of your own data as well as the relationship between your work and the findings of others (results and discussion may be combined).

6. *Conclusions*—a summary of your findings and their significance and perhaps suggestions for further research or applications for the findings

7. *Bibliography*—references or literature cited

8. *Appendices*—related materials that support a point and provide additional information but are not essential for understanding the thesis itself

9. *Abstract*—required for doctoral dissertations and may be needed for the master's thesis

Writing the thesis will be easier when you visualize a clear picture of the content and organization involved. As with your proposal, the content of your thesis will begin with presentation of a question to answer or a problem to solve. You will establish objectives for your study by reviewing the literature and suggesting a hypothesis for answering the question. If your proposal is carefully written and followed, it can become a foundation for the introduction, literature review, and methods sections of the thesis. Once you have tested your hypothesis with your methods, you can report and interpret results relative to the original question. With this vision of your thesis in mind, consider use of the following resources.

Graduate College Requirements

Most graduate schools furnish information on requirements in a catalog and in a guide for preparing theses and dissertations. Keep these instructions handy. There are deadlines to meet and fees to pay. Thesis requirements include time restraints, committee composition, and technical details such as margins, type face, spacing, and the kind of paper required. Knowing these things ahead of time will help you avoid problems later.

Style Sheets

For points of style beyond those specified by your department or graduate school, the discipline in which you are working probably has a style to which it generally adheres. For example, *Scientific Style and Format* (CBE, 1994) is a convenient reference on points of style such as abbreviations, punctuation, and bibliographic entries. If you don't know which style manual to consult, ask your advisor. Sometimes your advisor or committee will recommend that you choose a professional journal in your discipline and follow its style. Styles for publications differ with journal editors or publishers, but most provide instructions for contributors or other style sheets to follow. Become familiar with the style of your professional societies and have a style sheet handy. For further information on style, see Chapter 8.

The Library

The sooner you get acquainted with a library, the more time you will save. The simplicity or complexity of your literature search will depend on your knowing what you want to find and how to find it quickly. More information on the literature search and review is in Chapter 4.

Your Advisors

Your major professor is probably your most valuable resource. Take advantage of his or her expertise. Report to that advisor regularly, but don't make a nuisance of yourself with questions that can be readily answered with a graduate school catalog or your style manual. Because departments and graduate divisions differ in their requirements for theses, I cannot provide the final word for what your thesis should be. Your major professor can. In addition, your other committee members can be valuable consultants as you proceed with your study. Each is on your committee for a particular reason. Get acquainted with them early, and visit with them periodically.

Other Professionals

Unless you and your committee members have clear expertise in all areas of your project, you may need to consult specialists in addition to your major professor and the committee. For example, your thesis will most likely contain quantitative data, and a statistician should be consulted before you plan your experiment and collect the data. Be able to furnish him or her with an outline or thesis proposal that includes your research hypothesis and objectives. Formulating the design for your experimentation and stabilizing your plans early in your research can pay big dividends in both time and research quality later. You may also have available a writing center or someone with expertise in writing, revising, and editing. These services may prove valuable in getting the thesis put together well.

AVOIDING PROBLEMS

Your thesis should be the written record of your graduate research project and contribute substantially to your professional reputation. It will probably form the basis for a final graduate defense or oral exam by your committee. Building and maintaining your reputation with your peers and faculty will depend on not only how good the final product is but also how you handle problems along the way. The proposal, your research, cooperation with others, and the thesis will illustrate your scientific ability and professionalism. How well you integrate these individual activities will determine, in large part, how successfully and quickly you complete your degree. Weeks, months, and even years of delay can result from poor planning and execution of the graduate research project and thesis writing. You may want to read *Getting What You Came For*, particularly Chapters 17, 18, and 19 (Peters, 1997) before you get far into your program.

Work closely with your major professor but assume full responsibility for your program. Don't wait for the professor to tell you to write a proposal, search the literature, and write a literature review. If these chores are not required by your department, do them anyway. If the proposal is not required, write one and take at least an outline of your plans to your professor and ask for his or her advice. Then suggest that you submit the entire proposal to your committee and meet with them for their opinions. Your major professor will likely be pleased that you are assuming responsibility.

Work both cooperatively and independently, but don't get so independent that you step across the line of diplomacy and discretion. The department has policies, and you are probably using resources that belong to your department and your major professor. Consult with your advisor before taking drastic steps. As you become acquainted with departmental procedures and the personalities you work with, you will be able to determine how much independence you have. The following suggestions can help you avoid pitfalls common to graduate students.

Get Started Early

The responsibility for getting the thesis finished is yours alone. From the day you begin a graduate program, planning for your thesis begins (see Appendix 6, Fig. A6-1). Decide very early what area you want to work with so that your advisor, your course work, your exploration in the library, and your research in the laboratory or field can be chosen with the specific objectives of your thesis in mind.

Maintain Professional Relationships with Your Advisors

Recognize that advisors are humans with unique personalities. You are probably not the primary focus in their work, but they should be actively involved with your project. Don't be offended when they don't let your project take precedence over their many other activities. If there are times when your major professor is too preoccupied to help you, take charge of your own destiny, with finesse and discretion of course. Work cooperatively, but do some independent thinking too. That is part of being in graduate school. If you assume full responsibility for your program, you should finish your degree on schedule regardless of how much or how little input the advisor contributes.

Professors are abused in two ways: You ask too much or you ask too little. No professor has time for you. You waste their time with questions you could answer by consulting a dictionary, a college catalog, or a style manual. You waste their time with bits and pieces of your thesis that are too hard to read. You waste their time with too much casual talk or too many intrusions into their work schedule. However, a casual question or remark, a handy reference you find at the library, or a personal revelation—these things can be very important to the advisor's knowing you and your subject. Be discreet in how much time you take.

Graduate students also abuse their professors by avoiding them. Paradoxically, they may have plenty of time to work with you. Maintain a constant, congenial, and professional relationship with them. Your work is important to their own. Consultations on courses they teach, subjects of common interest, and your thesis—all are important for both of you, and advisors do not feel that such things are a waste of time. Discuss with them such issues as your time line, your responsibilities, and your position as author of any subsequent publications that come from your research. Certainly the research for your thesis and the writing of it are points that bear repeated discussions. A student who writes a mediocre thesis and dumps an error-ridden copy onto an advisor's desk 3 weeks before expecting to graduate is certainly abusing the advisor and jeopardizing graduation.

Draw Up a Carefully Planned and Well-Written Proposal

An attempt to pursue graduate research and produce a thesis without a written proposal is like taking a trip through unfamiliar territory without a road map. You may find your way and even be successful, but most likely you will waste time, have to double back over some roads, go down blind alleys, and even get lost. Working out hypotheses, objectives, justification, literature, and methods for the proposal will sharpen your perception of your subject. The written proposal provides an early draft or outline of the thesis, and it will serve both you and your graduate committee in communicating and keeping on track.

Maintain Accurate Complete Data

All data you collect should be considered important. Gather them carefully, write them legibly, analyze them thoroughly, respect their revelations, and then store all of them, not just the part you use. Read Macrina (2000) on keeping scientific records. Don't trust your memory. Write down field plans. Write down chemical analyses. Write down techniques and amounts you use. In field observations, note (write down) the weather. Record full references from the library with page numbers and full names (using only *et al.* can get you in trouble). Your good memory cannot always fill in the blanks.

In addition, for many of you a camera can help record data. Pictures you take can be important in showing results of your research and can often be used in the thesis and in slide and poster presentations. For these formats, you will likely want color photos. Keep in mind that at present, most journals publish only black-and-white photographs. You may need to record some information on black-and-white film. Images on most color film can be reproduced in black and white, but quality may be lacking. Digital photographs can transfer easily into your papers and presentations. Have access to a camera and use it often.

Write the Thesis as Your Work Progresses

It would be a formidable task to write the whole thesis at one time. But you can divide the work into logical portions. Both Peters (1997) and Bolker (1998) suggest that you write some every day. I agree. If you have not developed the capability of composing at a keyboard, now is a good time. Don't bypass the advantages offered by computer processing of words and data (Fig. A6-1).

FIGURE 6-1

Don't bypass the advantages offered by the computer for processing words and data.

Writing, designing tables and graphs, revising, editing, and printing can all be made easier with the word processor.

Before you begin your research or when you write the proposal, you should make a complete library search and write a rough draft of the **literature review.** At this point you can compile a first version of your **literature cited.** When you set up your proposal and plans for research, you can also write the **materials and methods.** As portions of your research are finished, you can draft the **results and discussion** section(s). All of these sections will require revision, but the easiest time to write each first draft is when details are fresh in your mind. Your **introduction** and **conclusions** can be the last sections you write. The introduction may be a modification of the introduction to the proposal. The purpose for the introduction is to direct the reader into the thesis; the conclusions need to focus the reader's attention on the most important findings. You can best accomplish these purposes after you see where you have been.

Be Proud of the Final Copy

Because it is a reflection of your reputation, you should not skimp on materials or quality for your final copy. You may have to pay for such services as typing, making figures, producing photographs, and providing 100% bond paper. These things can be costly; be alert to how to cut costs, but be sure the final product is a reflection you want. Others may contribute to it, but your name stands alone as the responsible author.

Finish Before You Go

The last weeks of your graduate program can be hectic. In addition to getting the thesis in a final form and completing other requirements such as a defense and a departmental seminar, you probably will also be looking for a job. Job offers may come before your thesis is ready to turn in. If you accept a position, arrange to begin work only after the thesis is complete. The new job will require your full attention, and even if you have some time in the evenings, you will find it hard to focus on the thesis. New graduates are often establishing a new home and family as they assume responsibilities for a new job. With your mind on the job and your new living situation, the thesis is not at the forefront of your consciousness. The psychological and academic influences of being near your major professor, the library, and other graduate students are important to your maintaining a focus on the thesis. Former students who have tried taking the incomplete thesis to a job repeatedly implore me to advise other students not to do the same. Finish before you assume the next job.

To Publish is to Build Your Reputation

The best time to publish is when your research and data are fresh in your mind. Sometimes a doctoral dissertation and occasionally a master's thesis can be published as a book or monograph. With or without direct publication, you may want to copyright your work. You can do that yourself or you may operate through a publisher. If your work is a doctoral dissertation, your graduate school will probably require the publication of an abstract with Dissertation Services (UMI Dissertation Services, 300 North Zeeb Road, Ann Arbor, MI 48106-1346). The graduate school office will probably take care of submitting the abstract for publication, but you may want to investigate other publication services that Dissertation Services offers such as on-demand publishing, obtaining copyrights, and binding.

Whether or not you copyright or publish your entire thesis, you should consider publication of your research results as journal articles. From your thesis, you can choose the most significant data and arrange them to agree with the format of a given journal. It is relatively easy to put a paper together while the data are new. If you postpone this effort, you often never carry it out simply because the task becomes increasingly difficult as time passes. You will also have new goals to accomplish that do not allow time for going back to old data. Without extension of your materials into an active journal, much of your valuable data can grow stagnant in your thesis on a library shelf. Some advisors now suggest or may even require that the thesis itself be styled to the form of journal manuscripts. Seriously consider this possibility and work toward it from the beginning of your program.

Last-Minute Jobs

After your thesis is written to your own satisfaction, final chores will likely take several weeks. In preparing for your final committee meeting or defense of your thesis, be sure to get copies to committee members at least 2 to 3 weeks in advance to give them plenty of time to read and evaluate your written work. You may also want to visit with each of them individually for specific suggestions before the final meeting. New suggestions invariably come from individual committee members and their joint considerations. Allow time after their reviews to change entire sections and particular details.

When you have polished your thesis by incorporating the suggestions from committee members, typically you should take your letter-perfect thesis to them, along with three or more copies of a signature page for final approval. You may be lucky. You may get the signatures in 2 hours. However, 5 days later you may be still trying; one committee member is out of town collecting research specimens, another is at a meeting in New Orleans, and a third just happens to be out of the office every time you check. Still another committee member would like time to look over the revised thesis before signing it. That person is not being disagreeable or picking on you. Advisors should look at what they are signing. Their reputations, as well as yours, are at stake. Although almost any student wishes for quick approval of the final thesis, serious students know that their own degrees and reputations are far more highly respected if permissive is not the key word in the reputation of the department from which they graduate. The professor should check your thesis a final time and even ask for further alterations in text, if needed, before signing the approval page.

The last minute has now evolved into weeks, and you still are not finished. Double-check your graduate school instructions for special requirements such as paper quality and margins. Final copies must be collated with no page missing. Even the best equipment or printing service sometimes errs. When you deliver copies to your department, the library, or the graduate office, your thesis can be rejected if you have not been careful enough with details.

Along with the thesis you may need copies of a signed copyright release to allow library personnel to reproduce your work for research purposes. Because this signature must be yours, don't leave your thesis with a friend to hand in without signing the forms. Obtaining original signatures, even your own, can take days and cancel your graduation if you are pushing deadlines to their outer limits. With your thesis delivered to the designated offices, you are almost finished. You should offer a copy to your major professor and even to committee members who have special interest in your subject. Keep a copy for yourself. You will probably not want it in sight for some time, but someday you may even display it—at least on a bookshelf at your home or your office.

PLANNING THE THESIS

A first step in planning your thesis is to determine the **format** you will use. Your choice of form partly depends on the content of the thesis and whether or not the work can be published as a monograph or as journal articles. The choice also depends on what form is acceptable to your graduate school, department, and major professor.

In reference to doctoral dissertations, the Council of Graduate Schools (1991) found that the "traditional" thesis is "alive and well at all universities participating" in their study and that this form "continues to represent the model in all fields." The council recommends "flexibility with respect to form." However, the "consensus is that simply binding reprints or collections of publications together would not be acceptable as a dissertation." Dissertations "require a fuller review of the relevant literature and a more complete discussion of results and conclusions than a journal would allow."

Gaffney (1994) has edited a policy statement by the Council of Graduate Schools on the master's degree, but that statement has little to say about the thesis itself. Requirements for the master's degree differ even more than do those for doctoral programs. Some schools do not require a thesis, and others involve other capstone experiences such as a performance, extra hours of specialized course work, or other special projects. In the sciences the thesis is often preferred. What the council does say about the master of science theses indicates that we can adapt standards for the doctoral thesis to the master's level. The council contends, "A master's student who does a thesis or project should be required to design the research project with the help of a faculty advisory committee, conduct the necessary background literature search, do the research, analyze the results, write the thesis, and communicate the results at an oral thesis defense." Unlike the requirement for a doctoral dissertation, the council suggests that the master's work "will not necessarily be original research, but it will be a new application of ideas."

These standards set forth by the Council of Graduate Schools do not designate forms acceptable for either the master's or doctoral thesis. Departments in the life sciences commonly accept, or some even prefer, the incorporation of journal manuscripts into both theses and dissertations, but the quality and standards established with the traditional thesis should not be sacrificed at either level. Substituting publishable journal manuscripts for a traditional thesis or dissertation offers the distinct advantages of providing the graduate student with experience in writing for publication and of publishing concurrently with acquiring the degree. Although The Council of Graduate Schools (1991) contends that the journal manuscript alone does not fully encompass the traditional requirements for a graduate thesis, literature reviews and complete reports on data collected can be incorporated into a thesis along with review-ready or published journal manuscripts. You need to understand what is expected and how to arrange the supplementary materials with the journal manuscripts.

ACCEPTABLE FORMATS FOR THESES

What then should be put before graduate committees and is acceptable for a master's thesis or doctoral dissertation? Because that question has no specific answer, your job is to decide what form you prefer and be sure your graduate school and advisors will approve your choice. What I describe here may or may not be acceptable to your committee.

The same basic outlines for traditional theses or those incorporating journal manuscripts will fit both the master's and doctoral levels if a clear understanding exists regarding the differences and the educational requirements for each.

Consult with your major professor before you decide which format to pursue. Consider the following possibilities.

The Traditional Thesis or Dissertation

Basic Outline

Introduction

> May be combined

Literature review

Materials and methods

Results

> May be combined

Discussion

Conclusions

Bibliography

Appendices

Samples of well-done traditional theses are available in any university library. Until the 1970s, this form for the thesis was expected from almost all degree candidates, and you and your major professor may still prefer it. Also, with exploratory research suitable for a master's degree but with data inadequate for publication, we can simply use this model and require a thorough literature review, a clear presentation of all procedures and results, and perhaps suggestions for additional research that may ultimately lead to publishable information.

We expect that the traditional thesis will follow the outline above, but reasonable creative alterations are not only acceptable but often commendable. Some schools may recommend length, but keep in mind that, unless a minimum length is required, for any thesis format you choose, the length of the thesis is essentially irrelevant. Some studies will need a 200-page report; others would be padded if they went far beyond 50 pages.

Theses or Dissertations Containing Journal Manuscripts

The thesis styled with journal articles in mind will be a compilation of manuscripts, which can be excerpted to be published, as well as sections such as the literature review and, perhaps, conclusions that are important to your entire study. If this thesis format is planned from the beginning of your study, much of the writing can be done before your research is complete. The literature review and its bibliography will need only to be updated at the conclusion of your program. Each journal article can be partially written as you recognize which objectives and methods will be included.

Basic Outline

Introduction

　　　　　　　　　　　⟩ **May be combined**

Literature review

Journal article I—style of a specific journal

　　Abstract

　　Introduction

　　Materials and methods

　　Results and discussion

　　Literature cited

　　Tables and figures

Journal articles (*ad infinitum*) journal style

Complementary studies—for auxiliary, preliminary, or supplementary

　　work not suitable for journal publication

Conclusions

Appendices

The master's thesis, including journal articles. For the master's candidate who has produced findings for a journal article but has acquired supplemental material that should be recorded in a thesis, I suggest the following possibilities. (The thesis will not necessarily be written nor arranged in this order.)

1. Write a complete review of the literature when your proposal is designed and before your research is done. You will update this review as you end your degree program, and it will become a separate chapter in your thesis.

2. After your study is complete or as you complete the analysis for data sets, write the journal manuscripts and note what you have omitted from those. You may have one publishable manuscript or several, but you probably will have material that is not publishable.

3. Write a complementary section on auxiliary studies or supplemental data collected that are not included in the journal manuscripts. This portion of the thesis may appear in the form of a manuscript you do not expect to publish, as a report or series of reports, or as appendices. The form depends on what remains to be presented.

4. If appropriate, write comprehensive conclusions to your entire study. If the complementary section is included as appendices, the conclusions may precede that section. These conclusions will be commentary on the extent to which you have accomplished your objectives and satisfied the initial hypothesis. They may contain suggestions for further research and suggested applications for your findings. Theses will differ a great deal in how extensive the conclusion section should be, and with discussion in and conclusions to other sections, some theses may not benefit from additional conclusions.

5. Rather than a complete bibliography at the end of the thesis, place references or literature cited at the end of each chapter. At the end of the literature review, you will have a list of all citations contained therein. Literature with each journal manuscript will be presented with the respective chapters in the style for the designated journal. This technique requires duplication of citations from the various chapters, but it allows for review-ready copy for the manuscript. Similarly, numbering of tables and figures can be by chapter rather than sequentially throughout the thesis.

Doctoral dissertation incorporating journal articles. The doctoral candidate must do all the master's student has done and more because original research and creative ideas are expected. The doctoral dissertation is likely to be made up of more than one journal manuscript and to contain additional supplementary materials. If the philosophic requirements for your degree are to be met, you must present a more extensive conclusion with commentary on scientific theory and philosophy that could not be published in journal articles. In addition to the same basic suggestions proposed for the master's thesis, I recommend the following for the doctoral dissertation.

1. Even after you have written two or more journal manuscripts, you will have a great deal of supplementary material left over. You may need to consider presenting those data in more than one chapter as well as in appendices.

2. You will review the literature more extensively and include ideas from related published works that may not have direct reference to your own objectives but do support the scientific principles behind your objectives.

3. As with the review of literature, the conclusions section to your thesis should treat your subject more intensively and extensively than would the master's thesis. Discussion and conclusions will demonstrate an ability to think creatively and to integrate scientific ideas from the literature with your original research.

In other words, your thesis as a doctoral candidate is a *dissertation*, namely, a lengthy treatise dealing with a subject from several perspectives that can direct your own scientific thinking and stimulate that of others in order to make a significant contribution to the scientific community. It is a document that helps to demonstrate your qualifications for the doctor of philosophy degree.

Caution. If you publish articles before the thesis is complete that you intend to include in your thesis, you may have to transfer copyright privileges to the publisher. Be sure to reach an agreement with that publisher to give you the right to publish the same material in your thesis.

Theses Containing Journal Manuscripts and a Proposal

Although I have not found it to be a common practice, a thesis or dissertation might also be produced with the inclusion of journal manuscripts along with a proposal. The proposal would be an initial chapter in the thesis and might contain the comprehensive literature review. More likely the literature review would be a separate section coming before the proposal section. These sections would be followed by the journal manuscripts and whatever complementary chapters, conclusions, and appendices were necessary. Except for the proposal and possibly the literature review, the outline for such a thesis would be the same as that for the thesis containing the journal manuscript.

One concern with this form for the thesis is that the proposal may need to be modified as the research progresses. Almost any extensive graduate study must evolve from an original plan, but new discoveries may be made or new methods discovered that require you to deviate from the original proposal. Because the proposal is needed to establish the course of study, changes made along the way need not discredit the original plan or create a need for rewriting it. A record of the entire graduate program can be seen in the thesis that begins with a proposal, is carried through reports on the research, and ends with a discussion of the successes and failures and their significance.

Background information on scientific theses is good in Stock (1985), Smith (1998), and O'Connor (1991). Bolker (1998) and Peters (1997) have practical suggestions for getting the thesis written. Many of the recent references that focus solely on theses have emanated from the humanities and social sciences and essentially ignore the unique characteristics of the scientific thesis. They do contain good general information, however. The one I'd recommend first is the guide by Mauch and Birch (1998). You might, however, enjoy reading the one by Fitzpatrick *et al.* (1998). That book is by three people who have just finished their dissertations. It is easy to read with light flavor and good humor, but it also deals with the serious issues.

Caution. Be sure that you have the approval of your advisor, committee, and graduate school before you decide on the form for your thesis. The ideas expressed here are not necessarily new and certainly not revolutionary, but many advisors or graduate schools have rigid criteria that may not be met with these suggestions.

THE THESIS DEFENSE

Most science departments probably require an oral exam or a defense of your thesis especially at the doctoral level. Although students worry and fret, this event is typically not confrontational and can even be enjoyable if you have done a good job with your research and your writing, have timed the situation well, and can relax. Timing is important. Be sure you get copies of your thesis to committee members 2 to 3 weeks before the defense. Attitudes developed by being rushed to read the document can be detrimental to your academic health during the event. You may also want to talk with individuals before the group defense to discuss particular issues that relate to their individual expertise. When your major advisor and you decide that the thesis is ready, distribute copies and schedule the defense. Make the time as convenient as possible for everyone concerned. Sometimes it is difficult to get all committee members together at the same time, especially if you wait until the last minute to schedule. If possible, schedule the meeting in a comfortable room or arrange chairs so that you are simply one of the group and the arrangement is not a physical one of "you against them." A round table arrangement is good.

Before the final defense, you may have made a presentation on your research to your entire department with all your committee members present. If not, you may want to make a short presentation as you begin the defense to introduce the main points and findings of your study. Your major professor can advise you on the feasibility of such a presentation. If you do one, be sure all equipment is working and the physical arrangement allows everyone to comfortably view your visual aids. Do a good job presenting. Although much of the decision on the worth of your study should rest with the written thesis, human beings (committee members are human) are always influenced by the skills of a presenter. After that presentation, you will simply join the group for their questions, your answers, and discussion.

Every committee is different and will treat the defense differently, but typically their goals are similar. Most are not there to harass you or to ask questions they should not expect you to be able to answer. Although they are evaluating your knowledge and the strength of your work, typically if you have succeeded to this point, you can pass this defense. Be conscious of the question in your audience's minds. As they read your thesis, they are asking such questions as: "Is the problem well defined here?" "How is it handled?" "Is the background literature accurate and appropriate?" "Are the results clear with data well presented and interpreted?" "Is the discussion meaningful?" "How thorough was this study?" "Is the thesis well written and easy to follow?" "Are the parts organized and written appropriately (e.g., title, abstract, methods, appendices)?" If they are satisfied with positive answers that they receive to those questions, the oral defense will probably be easy.

The questions they ask you during the defense will depend upon their reactions in reading the thesis and in having worked with you through your program. Of course, the questions they ask will differ for every research project, but the specific questions that you are asked will probably fall under these generalities: What do you believe are the strong and weak points in your study? What was the most important question that you answered in pursuing this research? What have you learned? How will you use what you've learned? Do you see any biases or unanswered questions remaining relative to your objectives? On the basis of your findings, what would you recommend for future research or application of your results? How does this work contribute to the discipline? What other work is currently being done in this area?

In addition to questions about your project and the science itself, you may be asked about the form and style of the thesis, findings in the literature, why you used certain methods, why you used tables or certain kinds of figures, why you chose certain statistical analyses, or where you will publish this information and why. Some final oral exams consist not only of a defense of the thesis but also questions that any master's or doctoral degree recipient should know about the discipline. If this kind of exam is given, be sure you are familiar with current events in your discipline as well as some history and the basic principles of the science behind your work. You can find out what kind of exam or defense you will have by talking with your advisor or other committee members and with graduate students who have experienced the event recently in your department. Peters (1997) has good suggestions in Chapter 19 about how to prepare for and perform during a defense.

References

Bolker, J. (1998). *Writing Your Dissertation in 15 Minutes a Day*. Henry Holt, New York.

Council of Biology Editors (CBE) (1994). *Scientific Style and Format: The CBE Manual for Authors, Editors, and Publishers*, 6th ed. Cambridge University Press, Cambridge.

Council of Graduate Schools (1991). *The Role and Nature of the Doctoral Dissertation: A Policy Statement*. Council of Graduate Schools, Washington, DC.

Fitzpatrick, J., Secrist, J., and Wright, D. J. (1998). *Secrets for a Successful Dissertation*. Sage, Thousand Oaks, CA.

Gaffney, N. A., ed. (1994). *Master's Education: A Guide for Faculty and Administrators*. Council of Graduate Schools, Washington, DC.

Macrina, F. L., ed. (2000). *Scientific Integrity: An Introductory Text with Cases*. ASM Press, Washington, DC.

Mauch, J. E., and Birch, J.W. (1998). *Guide to the Successful Thesis and Dissertation*, 4th ed. Marcel Dekker, New York.

O'Connor, M. (1991). *Writing Successfully in Science*. HarperCollins Academic, London.

Peters, R. L. (1997). *Getting What You Came For*, revised edition. Farrar, Straus & Giroux, New York.

Smith, R. V. (1998). *Graduate Research: A Guide for Students in the Sciences*. ISI Press, Philadelphia.

Stock, M. (1985). *A Practical Guide to Graduate Research*. McGraw-Hill, New York.

7

PUBLISHING IN
SCIENTIFIC JOURNALS

"Ich habe nichts dagegen wenn Sie langsam denken,
Herr Doktor, aber ich habe etwas dagegen wenn
Sie rascher publizieren als denken."
(I don't mind if you think slowly, Doctor; but I do mind
if you publish faster than you think.)

PAULI WOLFGANG

Almost all landmark publications, such as that of Watson and Crick on the structure of DNA, are the result of volumes of reports preliminary to putting that final piece into a scientific puzzle. That final piece often gets the publicity, but without the aid of past discoveries, the breakthrough would be impossible. Perhaps every scientist dreams of making a groundbreaking discovery in research and of publishing an article that will be considered the classic of the discipline. Not everyone can be that one-in-a-million scientist, but you certainly can provide reports that serve as the stepping stones toward that classic publication. The most common publication forum for written communication with other scientists is the scientific journal. Technology may be changing the forms for the communication, but the primary report remains essential for information transfer among scientists. Contributions to the journal literature can go far in building your professional reputation; however, not only must your discovery be new and valuable for others to read about but also the manuscript must be a well-written, clear disclosure of information.

85

When data are collected and analyzed and the results of your research are ready to publish, you have several decisions to make. When can you finish the writing? Will you have coauthors? Who will give you helpful reviews? Which journal will you submit your manuscript to? How soon will it be published? What is the possibility for acceptance? How do you deal with editors? Most of these questions do not have simple answers, but some general ideas may help you to follow the process of writing for journal publication.

PLANNING AND WRITING THE PAPER

Before you write the paper, determine who your coauthors will be and your position with them. Read about giving credit in *On Being a Scientist* (Committee on Science, Engineering, and Public Policy, 1995), and check what Bishop (1984) or style manuals such as the *CBE Style Manual* (1983) have to say ethical responsibilities for authorship and multiple authorship. Each author should have made a real contribution to the research, should be responsible for content of the manuscript, and should be involved in writing and reviewing the paper. Also, discuss with any coauthors the order in which the names will occur on the paper.

You and your coauthors should select the journal to which you will submit your manuscript before you write or by the time you have a rough draft written. The best journal or audience for one manuscript is not the best for another. Scan article titles to be sure you know what subjects are accepted by the journal. Read a few articles closely and examine their quality, style, and subject matter. Determine whether the paper will be refereed and who the publisher is. Papers that are refereed, or reviewed, are almost always better because of this process. You want your publication to be in good company; the reputation of the publisher could reflect on your own. Your own professional society probably publishes quality journals, and some other publishers are equally reputable.

When you have decided which journal to submit your manuscript to, acquire the "Instructions to Authors" from an issue of the journal, by writing to the editor, or from the journal's website. Many journals publish guidelines in at least one issue annually, and some have instructions in each issue. When you have studied the journal and its guidelines, write your first draft with the audience and the publication style of the specific journal in mind. If you later decide to submit to another journal, again study the publication guidelines and revise your paper to its style.

In selecting the journal to which to submit your work, you should consider circulation and the probable interval between submission and publication. Some journals are distributed worldwide, and the articles may be abstracted by national or international services and databanks. The subject of your particular manuscript may be better for a local, regional, or national publication. Many journals provide information in a footnote on the time from the receipt of the initial manuscript to its publication. This period can be measured at least in months, and often publication is well over a year or more from the time the editor receives the manuscript. Some publications are known for a rapid turn-around time, but they may or may not be refereed or known for discrimination in the manuscripts they accept.

Write the paper as you do the research. Now is the time to make good use of O'Connor (1991) or follow Day's (1998) recipe for writing a scientific paper for publication. Get background literature together and write a working title and a rough draft of the introduction before your results are available. Write a preliminary abstract without including results. This abstract will help to keep the justification, objectives, and main point in your mind when you begin to consider the results. Write the materials and methods section when you set up the experiment. Then, when results are ready, you can write the results and discussion section(s), the conclusion, and the revised introduction and title. Finally, revise the abstract by inserting results and a concluding remark and paring the whole thing down to size. Remember, every section of the paper will have to be revised several times before it is ready for publication. As a graduate student, you may be producing a chapter for your thesis as well as a manuscript for submission to a journal. Make whatever adjustments you need for the two documents.

AFTER THE PAPER IS WRITTEN

When you have written and rewritten your paper and every coauthor has reviewed and revised it, you will reach a point at which you cannot see how to make the communication any better. It's then time to ask for reviews from your peers, professors, or other colleagues. Most institutions or companies you work for and certainly the department in which you are doing graduate work will suggest or require that you obtain in-house reviews before submitting the paper to a journal. Many will require approval before submission and publication. Know the requirements of your department or employer before you submit a paper to a journal.

Whether or not in-house reviews are required, get opinions from colleagues, other than the coauthors, before you submit your paper for journal reviews. Choose your in-house reviewers carefully. It is good to have two or three. One may be someone who is very familiar with your work; he or she may see something important that you had assumed was obvious and failed to include in the text. Choose a second reviewer who knows nothing or very little about what you have been working on but has expertise in similar scientific matters. This reviewer can best give an objective look at both the science and the communication. You may wish to ask for at least a partial review from someone with skills in a special area such as statistical analysis. In addition, an editor or other reviewer who may not be a scientist can often improve readability and organization.

Finally, with in-house reviews complete and with further revisions of the paper based upon these reviews, you are ready to submit your paper to the journal. Read again the instructions to contributors or other information that is available with the journal or from the publisher. That information will probably tell you about page charges and submission requirements for page size, line spacing and numbering, and other matters of form and style. Check all details carefully and perhaps consult the checklist in the *CBE Style Manual* (1983) before submitting the manuscript. It's easy to forget an important point.

Actually submitting the manuscript will depend on the publisher's preference. Many publishers require, prefer, or accept electronic submission. They will furnish you with guidelines for submitting the paper electronically through the Internet or by submitting a diskette. They may want a hard copy of the manuscript, and the editor may require three or more copies of the paper for reviews. Submit the original, except for figures or photographs, and send neat, clear copies. For figures and photographs, make clear copies and retain the originals until your paper has been accepted. You don't want a reviewer marking up your only original. Send them when the editor requests them, usually with a final version of your paper. Clear all these issues for submitting the manuscript by reading guidelines or talking with the editor.

Submit your paper to only one journal. Instructions for submission usually indicate that the paper will be considered only if no other journal is concurrently considering it. You may think your chances for acceptance are better if you try two or three journals, but the publishing staff and reviewers can hardly afford to spend time and money reviewing and editing your paper only to have it published by another journal. Wait for a rejection from the first journal or ask that your paper be released by the first editor before you submit to a second. If you believe the first publisher is taking too long or asking for revisions that you cannot make, ask that the paper be released and send it to another journal, but until then, be patient.

When submitting a paper, follow the editor's instructions for electronic submission, or with your hard copy include a cover letter requesting that the manuscript be considered for publication in that journal and giving the editor a phone number and an e-mail address as well as the mailing address where you can be reached. To indicate the suitability of the manuscript for the journal, in your letter you can note briefly which section of the journal it is best suited for or describe the kind of research findings you are reporting. Most editors will send an acknowledgment that the paper has been received and sent to review. If you do not receive such information in about 3 weeks, call the editor to confirm that the manuscript was received. Then wait. If after about 2 or 3 months you have no word from the editor, call again to ask about the status of the paper. It may be lying on an absent-minded reviewer's desk. Your phone call may remind the editor to check with reviewers.

Acceptance of your paper depends not only on how good the research is and how well you write but also on the suitability of the subject and the acceptance rate of the journal. Acceptance can be influenced by the number of submissions to the journal. The rate of acceptance by many journals ranges between 65% and 75% of the manuscripts received, but some prestigious ones accept less than 15%, and for others the rate may be more than 80%.

THE EDITING AND REVIEWING PROCESS

Editors are human beings. As humans, they have all kinds of personalities; they are sometimes amiable and sometimes surly. Deal with them the way you do other humans; most will appreciate direct, open communication with you and will work with you and reviewers to produce the best possible paper. Don't be afraid to discuss and reject editorial changes that could result in a focus or meaning in your paper that you do not want, but listen closely to any criticism. The editors probably have more experience in both writing and publishing than you have. But ultimately the paper is yours. What is published appears not under the editor or reviewer's name but under your name.

Editorial staffs are organized in different ways. The journal's instructions to authors may outline the review and editing process used. Editors usually log in a manuscript when it is received. If the format and the subject matter are appropriate for the particular journal, that editor will then either seek reviews or send it to an associate editor who will seek reviews and communicate with you. Review processes also differ, but probably at least two other persons will review your paper. On the basis of the reviews and the editor's opinion, your paper will be accepted or rejected.

Seldom will a paper be accepted with no revisions suggested. If this rare event does happen in your life, you will simply be notified that your paper is accepted and going to press. More likely, your paper will be accepted, but revisions will be required. In this situation, your editor will evaluate the reviews, form an opinion, and send recommendations for revisions to you. You certainly may disagree with reviewers, but you should justify in writing to the editor any refusal to accept a recommendation. See Chapter 9 for more on reviewing and

FIGURE 7-1
Don't despair when you get that rejection.

revising. You'll probably make most of the revisions suggested and send the new copy back to that editor. You may receive a second set of recommendations and have to revise yet again. But when he or she is satisfied that the paper is ready for publication, the editor will notify you and put the paper into the publication process.

If the reviewers and your editor believe the paper should be rejected, it will be returned to you. The editor will probably provide an explanation for why the paper has been rejected and recommend possible revision that would make the paper acceptable. Don't despair when you get that rejection; most of us have experienced the same (Fig. 7-1). You can certainly discuss the quality or subject matter in the paper with your editor, but don't argue about any recommendation or rejection. That's not professional behavior. If you believe your paper is worthy of publication, revise it again and resubmit it or try another journal.

The following reasons overlap, but one or more of them may explain why your manuscript was unacceptable for publication:

1. **The research was inappropriate for that journal or was poorly conceived and executed.**
2. **The manuscript was poorly written or did not follow the style of the journal.**
3. **The research results are inconclusive; you have insufficient data or erroneous interpretations.**
4. **Interpretation is missing or discussion is unwarranted.**
5. **The research is trivial or incomplete, or the information is not new or is repetitious of earlier publications.**
6. **You have too much material; the paper is too long or padded with unimportant data or discussion.**

The publication process for professional journals is not perfect. Poor research and poor writing are sometimes published or good research and writing rejected by reputable publishers. However, the process may be as effective as human error will allow. If you familiarize yourself with review process during consideration of your paper, you will be less frustrated with editors and reviewers and with the time required to publish. Reviewers and editors can be very helpful, and your paper will usually be better if you take their advice. If you do not agree with a direct suggestion, you still need to consider it and determine whether you should make some adjustment in the text. Reviewers, who are usually not professional writers and editors, often know something is wrong but cannot identify the exact problem. Take their criticism of a point as an indication that the writing at that point is not clear. Most editors and reviewers are trying to make your paper better, not destroy it, with their suggestions.

Once your paper is accepted, you will have a few final chores to do before you see it in print or on a computer monitor. If you submit a diskette with the final version, be sure the version on the diskette and the hard copy are exactly the same. You may also be sent a copy of your manuscript called a galley, or galley proof, to be proofread in the form set up for printing. Proofread this galley with great care and have at least one other person proof it. In addition, you may be asked to sign a sheet to give copyright privileges to the publisher. This signature may transfer your copyright to the publisher, or the publisher may simply be asking for permission to print and reprint your work. Read the fine print to determine whether you retain the copyright. A final chore may be payment of publication charges. Authors are usually charged a fee per page of publication. Your instructions to authors will probably tell you what these charges are.

Always consult a recent issue of the journal to which you submit your work and read the instructions to authors or other information about that publisher's process. The steps and actions I have described are generalized, and they will continue to change as publishers update their technology to suit their conventions. Many changes are taking place with electronic publication, and you may see your publication on an electronic monitor rather than in a bound journal. Lessen your frustrations by keeping up with what's going on. If you are interested in what has evolved in publication, you might start with Crawford *et al.* (1996). But keep in mind that the ease in electronic dissemination of information removes none of the responsibility from an author for submitting a concise, well-written report.

At least two outstanding books are available for detailed information on writing and publishing a scientific paper for a professional journal: Day's (1998) *How to Write and Publish a Scientific Paper* and O'Connor's (1991) *Writing Successfully in Science*. O'Connor is not as concise as Day, but her book is thereby able to include more details. Still valuable with instructions and checklists for writing a paper for publication is *Writing Scientific Papers in English* (O'Connor and Woodford, 1976) Also, do not write your first scientific paper for publication without consulting the journal's notes and a style sheet or manual that a particular publisher uses. If you are a graduate student, your major professor or other faculty who have published repeatedly can answer many of your questions. Then, consult the journal editor about any questions that these sources do not answer for a particular journal.

References

Bishop, C. T. (1984). *How to Edit a Scientific Journal*. ISI Press, Philadelphia.

Committee on Science, Engineering, and Public Policy (1995). *On Being a Scientist*, 2nd ed. National Academy of Sciences, National Academy Press, Washington, DC.

Council of Biology Editors (CBE) (1983). *CBE Style Manual*, 5th ed. CBE, Bethesda, MD.

Crawford, S. Y., Hurd, J. M., and Weller, A. C. (1996). "From Print to Electronic: The Transformation of Scientific Communication." *Information Today*, Medford, NJ.

Day, R. A. (1998). *How to Write and Publish a Scientific Paper*, 5th ed. Oryx Press, Phoenix, AZ.

O'Connor, M. (1991). *Writing Successfully in Science*. HarperCollins Academic, London.

O'Connor, M., and Woodford, F. P. (1976). *Writing Scientific Papers in English*. Elsevier, Amsterdam.

8

STYLE AND ACCURACY IN THE FINAL DRAFT

"The simplest rule of thumb for the author and illustrator should be: follow what you see in print."

COUNCIL OF BIOLOGY EDITORS

Attention in early drafts of a paper should focus on organization and content, but you should also be conscious of format and details that are essential to the final draft. Before the paper goes to reviewers, you need to check carefully for accuracy and form in style and documentation. The final chore is to proofread carefully so that reviewers are not distracted by errors in style and grammar.

STYLE

In scientific writing when we refer to style, we are usually not referring to the effect that your personal creativity or literary training has on your writing. Rather we are interested in clarity and consistency in the use of words and other symbols. Style in this sense often refers to the typographical appearance and details in the form that your work takes. This **technical style** includes what is acceptable or unacceptable in organization, type font, capitalization, punctuation, indentions and spacing, abbreviations, citations and bibliographies, headings, footnotes, and any other conventions in the format or content of the entire work. Style is the way publishers or editors want to print your text to provide accuracy as well as a physical and verbal appearance to their publications and to your paper. The way words are physically arranged, capitalized, and punctuated can add clarity to the communication, but unfortunately, editors and publishers do not agree on standards for these details.

Despite efforts by the Council of Biology Editors (now known as the Council of Science Editors) and others, no consistent standards for style exist in the scientific community. And some traditional points of style have changed over the years, just as acceptable grammatical usage has changed. For example, you may have learned that numbers up to 10 should be spelled out in words and others written as numerals. Many journals are using all numerals except in specific instances such as at the beginning of a sentence. Similarly, in many scientific publications, the Latin notation *et al.* has been used to indicate more than two authors of a publication; however, the Council of Science Editors recommends the use of "and others" rather than *et al.* Some editors have adopted that recommendation; others have not. Almost all publishers of scientific journals follow similar organizational patterns, but the headings will differ. What one calls methods, another may call procedures, and although one may want you to combine the results and discussion sections, another may separate the two. Notice also the position and type styles used in headings and small details such as abbreviations and punctuation. Study carefully the style of the publisher to whom you intend to submit your work. Editors or reviewers are much more likely to read your paper with a positive attitude if you have been attentive to their style.

Although styles differ widely, some efforts toward uniformity have been made in recent years. The international system for units of measure is an important step forward. The style guide *Scientific Style and Format* (CBE, 1994) was produced with a goal of providing uniformity in technical style for all the sciences. (This style manual is being updated; check for a revised edition.) Other style manuals (e.g., CBE, 1983; Dodd, 1997) provide more discussion on issues such as planning, writing, and submitting the paper for publication, but they are less comprehensive on stylistic details. Most large professional societies and agencies print more limited style manuals for their contributors (e.g., American Physiological Association, American Medical Association, U.S. Geological Survey, U.S. Government Printing Office). A list of style manuals is provided in *Scientific Style and Format* (CBE, 1994). In addition, each scientific journal publishes an abbreviated style sheet called "Notes to Authors," "Suggestions to Contributors," or a similar title.

Decide where you wish to publish early in your writing and carefully follow the style sheet for that publisher. Failure to do so may mean that an editor will reject the manuscript even before he or she sends it to review. The title page near the front of the journal usually contains documentation about the publication and includes information about such things as subscription rates and publication charges as well as where to find information on the journal style.

STYLES IN HEADINGS

Manuscripts for journal articles, chapters in books and theses, proposals, or other documents or publications often need to be subdivided into sections. Headings provide transitional guideposts to link these divisions. Check a style sheet or an example of a publication to determine spacing as well as heading and subheading patterns, or make your own style neat, consistent, and attractive. Most important is to be consistent in print style, size, and position for headings that are parallel or of equal importance under a topic. Headings are usually designated as primary, secondary, and tertiary heads or first-, second-, and third-level headings. You will seldom if ever need fourth- or fifth-level headings. If a style sheet does not require that you follow specific patterns for headings, you may wish to use some of the following. Notice the symbolic emphasis given to the various forms. See Chapter 14 for more information on symbolic language. Positioning, spacing, boldface, underlining, italics, capitalization, or other symbolic dimensions can be manipulated to provide more or less emphasis to a heading. Determine which of these your publisher uses, and be consistent with parallel headings. I have attempted to arrange the following headings relative to their importance, but various forms of enhancement can change this order.

Sample headings

1. Center
   A. Underline and/or boldface

<u>Results</u> <u>and</u> <u>Discussion</u>

   B. Use total capitals
   RESULTS AND DISCUSSION or **RESULTS and DISCUSSION**
   C. With no added enhancement

Nitrogen Study, 2002

2. Left margin
   A. Above line, underlined, italic, or boldface
Nitrogen Study, 2002
   Text begins here …
   B. Above line, not embellished
Nitrogen Study, 2002:
   Text begins here …
   C. On line, italic, or boldface
Nitrogen Study, 2002. Text begins here …
   D. On line, paragraph indention, underlined, italic, or boldface
   Nitrogen Study, 2002. Text begins here …
   E. Pattern C or D with no embellishment

ACCURACY AND STYLE IN DOCUMENTATION

A most obvious point at which styles still differ widely among journals is in their techniques for handling the references. Both the form for citing the reference in the text as well as the form used in the bibliography differ from one journal to the next. A *bibliography* is a list of works related to a text. These works may or may not be literature (i.e., published) and may or may not be cited in the text. Even the heading for the bibliography differs between journals. Most of them are called "Literature Cited" or "References." The two terms may have slightly different meanings, but both are bibliographies limited to works cited in a text. *Literature cited* usually refers only to published literature that is cited in the text. And *references* may or may not be published forms but are cited in the text. Not all publishers adhere strictly to these definitions. Disagreement also exists about what is or is not considered literature or published.

The most essential criterion for documentation is that it be accurate and complete so that your reader can readily find any source of information that you have cited. Whatever the style of the publisher, the reference should include what is available of the following: **Author, date, title** (some journals omit this), and **source** (publisher and place of publication or journal name, volume number, and page). Every citation must be documented clearly and completely according to the style of your publisher. When a style for citations in the text and in the bibliographic list is not dictated by your publisher, at least be consistent, complete, and accurate. If you give erroneous details about a source, you will mislead and frustrate another researcher and degrade your own credibility.

Details in style differ in the reference lists for the various journals. Punctuation, capitalization, use of first names or initials and their placement — it is really amazing how many ways publishers can find to alter styles. It is easier to keep up with references if you use the name–year system as you write the paper even if you have to change to the number system or other styles when you write final drafts. The following are samples of some basic styles for references, but do not trust them entirely. Notice differences in punctuation, order, and other details. Look to the journal to which you are submitting your work; its style is sure to differ somewhat from any of my examples.

Three systems for textual citations and bibliographies

1. **Alphabet–Number System**
   Examples in the text:
   a. In 1988 Bilbrey and Rawls (2) developed a technique for....
   b. With the mathematical model (2), we could project....
   c. Several theories have been proposed for measuring soil water potential (2, 4, 7, 13, 21).

Advantages and disadvantages

This system provides little interruption to the text, is less costly to the publisher than is the name–year system, and can feature the name of the author(s) or a date in the text as in example *a* above. However, the number system does not always give immediate identification by name and year. If the writer adds or deletes a reference, he or she must carefully renumber citations throughout the text as well as in the bibliography.

The list of references is alphabetized and numbers correspond with the citations in the text.

Sample bibliography for the Alphabet–Number System:

Literature Cited

1. Adcock, R. L. 1998. Effect of moisture stress on soybean pod development. Crop J. 95: 345–347.
2. Bilbrey, J. C. and R. M. Rawls. 1988. Measuring soil water potential in a Sharkey clay. Gen. Soil Sci. J. 13: 121–124.
3. Green, C. R., A. C. Dobbins, V. C. Martin, and W. R. Amity. 2001. Response of grapes (*Vitis lubrusca*) to drip irrigation. HortReport 59: 13–14. (Online at http://www.fruitprod.net/waterneeds.html. Accessed 11 June 2003.)

2. Name–Year System:

Examples in text:

a. Bilbrey and Rawls (1988) developed a technique for....
b. By using a mathematical model (Bilbrey and Rawls, 1988), we could project....
c. Several theories have been proposed for measuring soil water potential (Adcock, 1987; Bilbrey and Rawls, 1988ab; Dobbins, 1991; Ferguson and Fox, 1999; Fox, 1991; Lennon et al., 1992; Watson et al., 1995). These theories allow for....

Advantages and disadvantages

With this system the writer can add or delete references easily. Some immediate identification of the reference is apparent. However, several such citations in a row can distract, and publishing cost is greater than for the number system.

The list of references is always alphabetized and may be numbered as well.

Sample bibliography for the Name–Year System

References

Adcock, R. L. Effect of moisture stress on soybean pod development. Crop J. 95: 342–345; 1987.

Bilbrey, J. C.; Rawls, R. M. Measuring soil water potential in a Sharkey clay. Gen. Soil Sci. J. 13: 117–121; 1988a.

Bilbrey, J. C.; Rawls, R. M. Drip irrigation for grapes. HortReport 78:14–15; 1988b.

3. Citation–Sequence System

Textual citations and bibliographic list are numbered in the order in which the citations occur in the text.

Example in the text

In 1995, Fox[1] developed a technique to measure soil water potential. Bilbrey and Rawls[2] modified that technique in 1988 to the form used today[3,4,5].

Advantages and disadvantages

This system is simple for printing short papers with short lists of references. A specific reference can be difficult to locate if the list is long.

The list is not alphabetized but is numbered and listed in the order of occurrence in the text.

Sample bibliography for the Sequence–Citation System
References
1. Fox, R. T. (1995) Agric. Bull. 102,47–49.
2. Bilbrey, J. C. and Rawls, R. M. (1988) HortReport 32, 17–21.
3. Lennon, T. R., Elzie, M. S., and Cola, R. C. (1992) Crop J. 79, 173–177.

In the preceding samples, all the examples are fictitious and various styles are used. These forms are all acceptable in some publications, but none may be appropriate to the editor and publisher to whom you submit your work. Always carefully follow the style of the publisher for whom you are writing.

Documentation of Electronic Sources

With more and more information being acquired from computer networks, clear citation to those sources is equally as important as reference to printed material. The important point in documenting material is that a reader can find the same source with the same information you retrieved. But standards for storing and documenting electronic sources are not yet stable. The material online is not archived as is printed text, and it can be changed or even removed from the network.

The same information as is used with printed sources should be included in the documentation of the electronic source: **Author, title, source** (publisher or journal), and **date of publication or the last update**. In addition, for the electronic source, list the **URL** (uniform resource locator) or access address and **the date you accessed it.** Printed sources remain the same; changes occur only by producing a new edition, which then also remains the same. For that reason, the date of publication is important. However, because information can be changed or deleted from the Internet, you should provide the date that you verified that it was there as well as the date it was last updated. Notice the example of an online reference under the Alphabet–Number System above, but keep in mind that specific editors may have other arrangements for citing that text. Problems in documentation of electronic sources are far from solved. Several guides are available on documentation of electronic sources, including Munger and Campbell (2001). Style manuals have some basic information on formats for electronic documentation, but when you need to use such citations, check to see how your journal or editor lists them. Be alert to new information on electronic documentation.

Headings and documentation are just two prominent areas in which styles differ markedly between publishers. You need to be conscious of other elements of style as well. Use of abbreviations, spacing, footnotes, and other inclusions will differ. With the capabilities of word processing and of electronic publishing, more and more of the responsibility for style will come to rest with the authors of a paper. Certainly you need to be conscious of the communicative value of how a page looks to a reader as well as what the words say. Above all be sure that you have proofread your final draft so that you do not burden the reader with all the distracting errors that slip by during composition. As C.C. Colton says, "That writer does the most, who gives his reader the *most* knowledge, and takes from him the *least* time."

PROOFREADING

Probably nothing blows smoke in the eyes of communication more quickly than the small mechanical error. Don't try to make early drafts of a paper perfect, but when you reach the final stages before sending a manuscript to your reviewers, then proofread as though the entire acceptance of the paper were at stake. It may well be. Readers come to your work from outside, and they may pay attention to content until they stumble across the misspelled word, the transposed letters, the word out of place. Then the mind jumps to that small point and can leave the content. The reader's mind is ticking with: "That's not right. Isn't it *i* before *e*? Shouldn't that be a capital? I never could spell that word either" and so on until what you were saying gets lost. Often people let important decisions about your reputation hang on the grammar and mechanics in your writing: "Look there. A misspelled word in the first line. Don't they teach spelling any more?" Before those readers are finished with reading, they may believe you are a careless, uneducated individual, and they will not hire you or publish your work. Perhaps I exaggerate the importance of accuracy, but I think not. Learn to proofread.

Proofreading is not editing. The proofreader is looking for the mechanical or grammatical error and only indirectly at the content. Proofreading is not like any other kind of reading. There is scanning, there is careful reading, there is studying texts for content, and there is speed reading. But none of those are proofreading. If I am asked to proofread a paper I am not familiar with, I proofread it at least three times. First, I read through the paper so that the basic content will not distract me the second time through. During that first reading, I may catch several errors that I check lightly. Then I read the text word by word, phrase by phrase, and scrutinize the innards of the words, the order of the phrases. As I find errors, I mark them lightly again. The third time I read with the flow of the words, sentence by sentence, paragraph by paragraph, yet again noting any error I missed in earlier readings. Then I go back and mark each error carefully with the appropriate proofreader's symbol. I am likely to read the text aloud a fourth time after it is marked to verify my marking and catch any error I missed.

Marginal

Correction	Symbol	Marked text	Corrected text
Delete	ℒ	the old cat	the cat
Restore deletion	stet	the old cat	the old cat
Close up space	⌒	the o ld cat	the old cat
Insert	o/d	the cat	the old cat
Replace	o/d	the big cat	the old cat
Insert space	#	the cat	the cat
New paragraph	¶	Once upon a time	Once upon a time
Move to left	⌐	⌐ the dog	the dog
Move to right	⌐	⌐ the dog	the dog
Center	ctr	⌐ the dog ⌐	the dog
Transpose	Tr	the on dog	on the dog
make lower case	/c	the Old dog	the old dog
Capitalize	cap	jim burns	Jim Burns
Period	⊙	Go west	Go west.
Comma	⌄	Come Jim.	Come, Jim.
Apostrophe	⌄	Jims dog	Jim's dog
Superscript	⌄	3 m2	3 m²
Subscript	⌃	H2O	H₂O

FIGURE 8-1
Common proofreader's marks

With word processing some proofreading can be done on the monitor and corrections can be made easily. Some people read from the screen more easily than do others. I recommend that at least one reading be done from hard copy. The screen may not tell you exactly how your work will look on the printed page. Use the spell check feature in word processing, but don't trust it implicitly. It misses real words that are used incorrectly, such as the substitution of *affect* for *effect*.

If you have trouble catching errors, read aloud or have someone read slowly to you while you follow the copy. Always have a second reader proof your final draft. It is most difficult to catch the errors in your own work. You know what you meant to say, and your mind will read in omissions or leave out any element that intrudes. Proofread and mark copy carefully. Figure 8-1 shows some handy symbols to know. Others are listed in unabridged dictionaries or style manuals.

As you write your proposal, thesis, or paper for publication, you are most concerned with the research itself and how to report that research to an audience. You must think about big issues of which data to use and how to organize and develop content for the greatest clarity. But, as Luellen (2001) says, "DO sweat the small stuff." Adhering to a consistent publication style and proofreading carefully can certainly influence the acceptance or rejection of your manuscript.

References

Council of Biology Editors (CBE) (1994). *Scientific Style and Format: The CBE Manual for Authors, Editors, and Publishers*, 6th ed. Cambridge University Press, Cambridge.

CBE. (1983). *CBE Style Manual*, 5th ed. CBE, Bethesda, MD.

Dodd, J. S., ed. (1997). *The ACS Style Guide: A Manual for Authors and Editors*, 2nd ed. American Chemical Society, Washington, DC.

Luellen, W. R. (2001). *Fine-Tuning Your Writing*. Wise Owl, Madison WI.

Munger, D., and Campbell, S. (2001). *Researching Online*, 4th ed. Longman, New York.

9

REVIEWING AND REVISING

"No one can make you feel inferior without your permission."

ELEANOR ROOSEVELT

"We are the products of editing, rather than authorship."

GEORGE WELD

For almost any craft or art form, the initial rough creation has to be reshaped, sanded, revised, polished, or fine-tuned as do the draft and subsequent forms of a scientific paper. In other words, good scientific communication is not the product of an initial attempt by a talented writer, but the end result of repeated reviewing and revising (Fig. 9-1). As Zinsser (1998) says, "If you find that writing is hard, it's because it is hard." Tough revision is essential, and good reviews can be helpful in this process. Few people write well; many people can revise well.

Recognize that a well-written paper will be reviewed and revised at least three to six times before it is acceptable to your audience. Many good papers are revised far more than six times before publication. A barrier to achieving a well-done final draft is a failure to admit that you, like everyone else, need the opinions of others and need to revise more than once. I assure you that no chapter in this book had less than six revisions, most of them far more, and that every review has been helpful. As a scientist, you will review and revise your own work, use reviews from others in revising your work, and serve as a reviewer for other authors.

FIGURE 9-1
Good communication has to be polished from an initial rough draft.

REVIEWING AND REVISING YOUR OWN PAPER

Learning to review and revise your own work requires an attitude of confidence but of objectivity and humility as well. To look at your own work objectively is difficult, but with an open mind, you can develop an ability to do so.

After you have written your first draft, carry your own reviews through at least three stages of revision before you expect someone else to work with the paper. You will first review and revise the general content and internal organization to determine that you are emphasizing the main point of the paper and that the order of the content is easy for a reader to follow. Then, revise individual parts of the paper to support the main point in the introduction, methods, results, supporting literature, tables and figures, and conclusions. Finally, pay attention to clarity and details in style, diction, and mechanics. Try some of the following ideas.

Lay your work aside for several days, or even 2 or 3 weeks, after you have written a good draft. If deadlines are pursuing you, at least take a break for a few hours. The psychological adjustment you can make by getting away from your own writing may save you time in the long run. When you pick up the paper again, try to read it as though you were completely unfamiliar with the research or the writing. Read through the paper without stopping to criticize any small points, just as though you were reading a published article by someone else. With the overall picture in mind, ask yourself whether "this author" emphasized the main point, whether he or she organized the material well, and whether too much or too little was said. In other words, adopt an outsider's attitude and criticize the general content. What questions would you have if you were not familiar with the work? Try to recognize whether the hypothesis, the objectives, the methods, and the results are clear to another scientist. If not, begin work on those points without worrying over details of sentence structure and grammar. At this point you are looking for general clarity, order, logic, and a focus on your point of emphasis.

When you have revised and are convinced that your main points are clear, read the revised version — this time much more slowly. Check content in the main sections of the paper: the abstract, the introduction, and so forth. Check through the materials and methods carefully. Could an outsider follow your instructions and produce the same experiment with the same results? Check your own data for accuracy. Look at the tables and figures and the text of the results and discussion. Can you simplify data for the outsider; do you need to add a point; or more likely, can you delete some discussion or even a column in a table or a whole figure without diluting a point? Check ideas in textual citations and in the references; consult the sources themselves for what they say. Have you distorted a citation with the context in which you refer to it? You are not going to do this sort of thing deliberately, of course, but it is easy to misinterpret an author or make an error in transferring details or ideas to your own words.

After you have revised the content of your paper in this way, read it again very slowly, phrase by phrase. This time watch the diction, small flaws in logic, and structure of the sentences. Does each statement support your hypothesis and your line of reasoning? Remember the works of Day (1998) or Tichy and Fourdrinier (1988) or Paradis and Zimmerman (2002), or Zinsser (1998). Are your sentences full of clutter? Rewrite them. Then, watch for precision and details. Check accuracy in spellings, numbers, and meanings of words; adhere to journal style. Don't trust your good memory or your intuition. Especially check data against original records, and verify all textual citations and references.

After you have gone through these three stages in the first revision, read the paper through again smoothly. **Read it aloud.** It is amazing what you will hear that you could not catch when you read silently. Is there something you missed in the revisions? You may need to repeat all three stages again, or the last two, or the last one. Most people would do well to put the paper aside again for several days and then go through the revision process step by step at least once more. At some point along the way, however, you should turn the criticism over to coauthors and in-house reviewers. Colleagues, supervisors, and editors can often serve as a buffer between you and the audience and can have very helpful suggestions. Almost all of us are myopic toward our own work and fail to recognize those points that an audience might not understand. Look to these in-house reviewers before sending it to the journal. Revise again after you receive their comments.

MAKING USE OF REVIEWERS' SUGGESTIONS

When you have revised according to your own and in-house reviews, then the paper is ready for submitting to the journal. Don't expect miracles from the publisher's reviewers. Although peer reviewers are helpful, they will not rewrite your paper for you or make a good paper out of a bad one. With most professional journals, the reviewers are volunteering their time. The only compensation they get is the satisfaction of participating in professional responsibilities. Allow them time to do a good job. Treat the reviewers with respect by first getting the paper to the point at which you think no further revisions are needed and by then giving serious attention to their critical observations.

When you review a paper for someone else, you should be candid with the author about what you see as strengths and weaknesses in the paper, and you should appreciate the same kind of candor for your own manuscripts. The reviewer has a perspective that is closer to the audience than that of the author. Take the review comments seriously and use them to your advantage. You need not agree with your reviewers, but you should respect their opinions.

Again, your attitude is important. Be prepared to revise yet again. Even though you submitted the paper fully believing that it was in need of no further revision, you will need to revise after the reviews. I have worked with hundreds of refereed papers. **One** among them was accepted for publication by a refereed journal with no revision requested. That one was a methods paper by a graduate student. Your paper is not likely to win such a lottery. The chances are remote.

If you expect to receive numerous suggestions for revision, you will be ready for what you receive. Read the reviews objectively. Generally, the reviewers are sincerely interested in helping to produce a good publication and are not trying to reject your paper. Approach all the reviews with this idea in mind, and try seeing your paper from their point of view. Your goal is to produce a better paper after incorporating or otherwise acting upon their suggestions.

Bad reviews can happen to good papers. And as Danielle Steele says: "A bad review is like baking a cake with all the best ingredients and having someone sit on it." Probably the worst reviews are the nonreviews or those that say very little. General comments such as "tighten the organization" or "well written; could add more data" are not very helpful. The reviewer needs to suggest points at which the organization is weak or you need more supporting data. Without rewriting the paper, reviewers can and should be specific.

Accept the fact that occasionally you will encounter reviewers who consider that only they understand your subject. They may play "king of the mountain" to try to push your research aside so that theirs remains on top. Don't let such people frustrate you; they can be unintentionally helpful. In their criticism, they will likely allude to the weakest parts of your paper, and even though you may not agree with their criticism, you can make the effort to strengthen that part of the manuscript. Any reviewer who finds something wrong, weak, or hard to comprehend is doing you and your paper a service no matter what his or her attitude is. The reviewer may not be able to pin down what is wrong with a part of your paper, but the fact that he or she is questioning it probably means it could be improved with rewriting. Maintain your own confidence in what you are doing, and let the criticism lead you to a better paper.

Be open-minded, but do not blindly follow suggestions by reviewers. They may be entirely wrong about a point. You may need to explain to an editor why you have not followed a reviewer's suggestion, but remember that the paper belongs to you and any coauthors who are working with you. Reviewers are people too; they make mistakes. Just don't conclude that the mistake is theirs until you have engaged in some objective self-criticism. Then revise your paper yet once again. Lay it aside a few days, and then review it yourself. When you believe the accuracy and clarity of the paper are pristine, return it to the editor. If each author, reviewer, and editor in the publication process performs his or her responsibilities in a professional manner, we can be proud of the scientific literature, including your paper.

REVIEWING THE WORK OF OTHERS

The most common encounter you will have with reviewing for others will probably be with manuscripts submitted to professional journals. Journal reviews can be full-blind reviews in which the authors' names are not revealed to the reviewer, and the reviewers' name is not revealed to the author. Or, at the other extreme, reviews are occasionally fully open; the author and the reviewer may not only be identified to each other but, along with the article, the journal may also publish the reviewer's comments and even a reply from the author. Many reviews fall somewhere between these two extremes; often the author does not know the reviewer, but the reviewer knows who the author is.

The peer review process for journal publication is certainly not foolproof, but it may be our best safeguard against too much inferior literature in the sciences. As a reviewer, reception of the paper by the audience is your chief concern, but you are in a position somewhere between the editor and the author. Try to understand whether your position is midway or skewed in one direction or the other. Sometimes the reviewer's major responsibility is to the author of the work to help get the manuscript ready for an editor; sometimes the major responsibility is to the editor to help judge and justify the acceptance or rejection of the paper and make specific recommendations to the author. Whatever your position, you should carefully **evaluate the work to the benefit of both the editor and the author on behalf of the audience.** Keep in mind that yours are mere suggestions. Both editing and rewriting are outside the realm of your responsibilities as a reviewer.

When you review a paper, three professional principles should be utmost in your mind. **One, do a good job.** Along with editors, reviewers are the guardians of the scientific literature; enough garbage slips into publication without your contributing to it by not giving due attention to a review. **Two, review in a timely manner.** Authors are anxious about their papers, and journals need to publish research before it gets too stale. Get your review back to the editor in the time requested. **Three, keep information in the manuscript confidential.** The work is the property of the author. You may run across exciting ideas in your reviews, but they are not yours, and public disclosure or your own use of the information is unethical until the work is published and then clearly referenced.

As you review a paper for someone else, you can follow the same three-step plan that I've outlined for reviewing your own paper. This time, however, you don't have the intervening revisions. You simply review and comment on major points, sections, and details from the same version. Keep in mind that your job as reviewer is not to revise but simply to make suggestions and let the authors act upon those suggestions as they see fit.

The reviewer's role is to peruse the entire contents with full attention to all details and accuracy in the communication. To be a full reviewer, you should be knowledgeable about the subject of the manuscript, you should have read much of the literature on the subject, and you should be familiar with the journal to which the manuscript is submitted and the audience who will be reading the published work. Only with this expertise can you judge overall quality in accuracy of scientific details and clarity in communication. Because the reviewer's role is one of assistance, you should maintain the attitude of humble but confident objectivity. The reviewer must express an opinion on the overall values of the manuscript and strengths and weaknesses in its component parts. The advice and check list for reviewers in the *CBE Style Manual* (1983) can be helpful to you in establishing your role and in developing an effective attitude. Dodd (1997) also provides a variety of opinions on peer reviewing. Appendix 7 is an excellent example of a review. The following suggestions may prove useful.

—Keep your own paper handy or make a photocopy of the manuscript to work with. Initially at least, do not write on the original manuscript. You may change your mind about your own opinions as you study the paper, or the editor or author may have requested that you not mark the copy.

—Read through the entire manuscript to gain a familiarity with it. Then read it carefully and consider the following questions: Is the research clearly justified and made credible with data and with appropriate allusions to scientific principles and the literature? Does the work add to the knowledge already available in a given area of study, or is it essentially a duplication of earlier publications?

—If you can give positive replies to those questions, consider the research itself as depicted in the paper: Is it valid, complete, and credible? Are the hypothesis and objectives clear? Is the statistical design appropriate, and are the methods presented in adequate detail for a reader to repeat the experimentation? Are the results clearly stated, and do the data support the interpretation given to the results? Are the data analyzed statistically, interpreted accurately, and presented in text, figures, or tables that are easily comprehensible?

—Now read carefully again and consider the manuscript by parts and in detail. Are the main points emphasized in the right way? Is the title appropriate? Is the abstract the right length, and does it contain the adequate details? Should some sections of the manuscript be deleted, revised, shortened, or expanded?

—Then, as in reviewing your own work, go to the smaller details. Is organization within paragraphs clear, and is the sentence construction effective? Does the author follow the proper format for the text, tables, figures, and references? Are all important references cited, and are any of them simply excessive baggage? You may even mark misspellings or other mechanical and grammatical problems that you see in the writing. But remember that your role as a reviewer is not to be a technical editor, and you should not rewrite for the author or edit for the editor. But do give specific suggestions about rewriting a point you find weak.

—You should be making notes on all these matters or marking a work copy as you read and reread the paper. Prepare your comments for the editor and the author on the manuscript itself or write a critique on a separate sheet. Write your comments clearly, with specific suggestions for how to make improvements. It doesn't mean much to an author or editor simply to say, "Not acceptable." *What* is not acceptable? Do you have a suggestion for making it acceptable? In other words, be specific without actually rewriting material.

—Finally, ask yourself whether you can support your criticisms. Have you been helpful or prejudiced? And have you communicated your opinions clearly to the editor and the author? Study the well-done review in Appendix 7. **A good reviewer is invaluable; a bad one is demoralizing and destructive.**

References

Council of Biology Editors (CBE) (1983). *CBE Style Manual*, 5th ed. CBE, Bethesda, MD.

Day, R. A. (1998). *How to Write and Publish a Scientific Paper*, 5th ed. Oryx Press, Phoenix, AZ.

Dodd, J. S., ed. (1997). *The ACS Style Guide: A Manual for Authors and Editors*, 2nd ed. American Chemical Society, Washington, DC.

Paradis, J. G., and Zimmerman, M. L. (2002). *The MIT Guide to Science and Engineering Communication*, 2nd ed. The MIT Press, Cambridge, MA.

Tichy, H. J., and Fourdrinier, S. (1988). *Effective Writing for Engineers, Managers, Scientists*, 2nd ed. John Wiley & Sons, New York.

Zinsser, W. (1998). *On Writing Well: The Classic Guide to Writing Non-Fiction*, 6th ed. Harper Collins, New York.

10
TITLES AND ABSTRACTS

"Clutter is the disease of American writing. We are a society strangling in unnecessary words, circular constructions, pompous frills and meaningless jargon…. Simplify, simplify."

WILLIAM ZINSSER

Titles and abstracts are the parts of your paper that will be read most often, and they may be the most difficult sections to compose effectively. What Zinsser (1998) says about clutter and simplicity is particularly true for titles and abstracts. They serve two purposes for your readers: (1) to disclose the basic information that the paper itself contains and (2) to help readers decide whether to read the entire paper. Together, the title and the abstract must help readers quickly identify literature that they want to read. As you write them, keep the purpose and audience clearly in mind, and keep your titles and abstracts as uncluttered and simple as possible.

TITLES

The title is the first impression you make on your audience. It should attract attention, but most important, it should be informative. The title may be the most notable phrase you write. Many people will read it, but few will read the rest of your paper. It should use:

1. **The most precise words possible**
2. **Words that indicate the main point of the paper**
3. **Words that lend themselves to indexing the subject**

One technique for creating a title is to write the objectives first; then write the rough title, sometimes called the *working title*. Go on to write the entire paper, and then rewrite the title. Write and revise the abstract, and then check the title again. It may need another revision.

The most common problems with titles perhaps are in their length and in the selection and arrangement of words. Be sure your title will make sense to someone not familiar with your subject. Use words that other readers might consult to find information such as your paper contains and use them in a sequence that is not ambiguous or misleading. The first source for key words for indexing services is the title. Study your title for unnecessary words and put the most important ones first. Provide adequate information, but keep your title relatively short. Eight to twelve words is a good range to work in. Scientific titles should not be newspaper headlines. Scientific readers are not looking for a journalistic sensation story; they want information. A full sentence with an active verb is usually not a good title. Just be informative; let the title tell what your paper is about.

The journal may also request a *running title*, or *running head*. This is simply an abbreviated form of the title that appears as a headnote on journal pages beyond the first page of an article. For more detailed information on titles, check Day (1998). No, **study** Day on titles; his remarks will give you a feel for what to do with your own title. Also, take a look at the versions of a title in Appendix 8.

ABSTRACTS

The term *abstract* is used loosely to refer to almost any brief account of a longer paper. Informative abstracts used with scientific journal articles are a more-structured form than this loose definition permits. With reference to *BIOSIS Guide to Abstracts* and to the *American National Standard for Writing Abstracts*, the *CBE Style Manual* (1983) distinguishes between the other brief forms and abstracts that are used with journal publications. What that manual describes as an abstract is called an *informative abstract* by Day (1998) and others, who compare it to a *descriptive abstract*. To complicate definitions further, some societies will request *extended abstracts* for publication in proceedings. These appear to be much shorter than a full paper but can contain more data than the informative abstract because the publisher is providing more space for the publication. Examine a previously published issue of proceedings to determine what is expected if your society requests an extended abstract.

The **descriptive abstract,** or indicative abstract, describes the contents of a paper but does not give a precise condensation of the information contained therein. Its contents would be relatively worthless if it were not accompanied by the report itself. It may be the best form for some reports and, similar to a table of contents, is helpful for a reader in deciding whether to read the entire paper. But one must read the entire paper for substance. Descriptive abstracts contain too little information and detail to substitute for the informative abstract that most refereed journals expect.

Don't let all this fuss over definitions misguide you. Just know that these strange breeds exist, and then recognize that for journal publications you need the **informative abstract.** Study the journal to which you will submit a manuscript to obtain specific instructions and to read examples of published abstracts. You will find a few differences between journals, but any informative abstract must serve several purposes:

1. To show the reader very quickly whether the full report is valuable for further study
2. To be extracted (abstracted) from the full report for separate publication
3. To furnish terminology to help in literature searches by individuals or by literature retrieval specialists in indexes and electronic databases

To serve these purposes, the abstract must be a short, concise, but completely self-explanatory report on a scientific investigation. Like the report itself, the abstract must include:

1. The research objectives and rationale for conducting the investigation
2. The basic methods used
3. The results and significant conclusions that can be drawn

Notice that the two parts generally included in the full paper that are omitted from the abstract are the literature review and discussion. A concluding statement may give an interpretation to the results, but any lengthy discussion or speculation is out of place. Although some journals limit the length even more, many style sheets specify that the abstract should not exceed 200 to 250 words or 3% to 5% of the length of the paper itself and that the form should be one paragraph. Even for less formal presentations such as papers presented at meetings or published in nontechnical forms, the abstract should still be a technical, concise, complete report. If such precision is out of place, an informal annotation or other summary should be substituted. Study the sample abstracts in Appendix 9. McMillan (2001) also presents and discusses some examples of abstracts.

Societies sometimes publish lists of what should be included in or omitted from an abstract. Let me try to hit upon some main points and refer you to the

CBE Style Manual (1983) and Day (1998) for what the abstract should include. Along with the essentials (i.e., research objectives, basic methods, results), the abstract should fully summarize the contents of the paper in as much detail and as few words as possible. In being concise, maintain clarity and avoid a telegraphic style that is hard to read. Because it is often read hurriedly in scanning the literature, it should flow smoothly. Emphasize the main points and avoid long lists of information. In presenting the main points, be as specific as possible. For example, say "20 and 40 kg ha^1 of nitrogen" and not just "two rates of nitrogen." Keep the tone strictly objective; avoid any loaded language that would suggest speculation. Provide any scientific information, such as scientific names for species, that is important for a complete understanding of your subject. Avoid abbreviations that are not immediately evident to the scientific community, and use no allusions to the literature or to any other material that would require a footnote or pursuit of external information.

When you think about it, all these do's and don't's for the abstract simply emphasize its purpose: to be a concise, complete report of your work that can stand alone without further explanation. In addition to the references listed here, study titles and abstracts in the journals in your discipline and any instructions your publisher provides.

References

Council of Biology Editors (CBE) (1983). *CBE Style Manual*, 5th ed. CBE, Bethesda, MD.

Day, R. A. (1998). *How to Write and Publish a Scientific Paper*, 5th ed. Oryx Press, Phoenix, AZ.

McMillan, V. E. (2001). *Writing Papers in the Biological Sciences*, 3rd ed. Bedford/St. Martin's, Boston.

Zinsser, W. (1998). *On Writing Well: The Classic Guide to Writing Non-Fiction*, 6th ed. HarperCollins, New York.

11

PRESENTING DATA

"Graphic excellence is that which gives to the viewer
the greatest number of ideas in the shortest time with the
least ink in the smallest place."

EDWARD R. TUFTE

Once you have analyzed the data from a research project, you are ready to put
them into a form to communicate to others. Before you decide what form your
data will take, consider your purpose in presenting the data and the audience
for which it is intended. Ask yourself what point you wish to make. Also, be sure
that you yourself understand the point, the data, and the data analysis and
firmly believe in what you are reporting. Then deliver the information with as
much clarity as possible for the medium you are using—a journal article, a
poster, a slide presentation, or other report.

You may be able to present data in the flow of the text. But often tables, graphs,
maps, photographs, flow charts, or other figures or illustrations can communicate more clearly than text can. It is not realistic to try to use all the data points
that you collect during your experimentation. As Gastel (1983) suggests, "Good
science communication, like good science, usually entails gathering much more
information than will appear in print." Your job as a writer or speaker is to select
representative data and to determine which form of presentation will be
most clear for the audience. Some audiences need a simple table or figure and
more verbal explanation, and they may not understand linear regressions or
logarithmic scales; others may be skeptical of your credibility if you do not use
these forms.

Fortunately, we have a wide selection of formats to use for data presentation. Once you have the audience in mind, think in terms of which format will best express your point. **Tables** are excellent for presenting specific data and making exact comparisons between data points. Tables can also show gradations and relationships between controls and treatments. **Bar charts** are not so numerically specific as tables but can make more dramatic comparisons. Bar charts make comparisons in sizes, magnitudes, and amounts, as well as other distinctions. They should emphasize differences rather than trends, but as with data in tables, the bars can be arranged to indicate a trend. Although measurements can be closely estimated relative to a scale on the axis, the purpose of a bar chart is not to show specific numbers, and seldom should you duplicate data by writing specific amounts above the bars. If you want the specific numbers, use a table. **Pie charts** can be a clear, simple way for showing parts of a whole or percentages. They are most effective when the pie has a limited number of slices. The smaller the slice, the more difficult it is to estimate the amount it represents. Slices can be labeled with amounts, but thin slices are hard to label. **Line graphs** are designed to demonstrate movement, change, and trends, especially over time or concentrations. Similar to bar charts, their purpose is not to show specific amounts, but they can show estimates against a scale. The semilogarithmic line graph can demonstrate relative changes in two values. Variations in tables, pie or bar charts, and line graphs give you a wide range of choices. Select the form that will make the data say what you want the audience to understand.

Results in a report are relatively easy to write or present once you have displayed your data in tables and figures and studied their meanings. Keep in mind the communication principle of simplicity. The simplest possible message is the most likely to be understood, but don't fragment the meaning by oversimplifying the data. Seldom if ever, do you need to use both a table and a figure to express the same information in written work. For a slide presentation, you may want to use both in order to move from one slide to another and give the audience two views of an important point while you talk about it.

In any communication, misunderstanding is a danger. It is imperative that you present your results with every degree of clarity and honesty possible. Of course, deliberately loading data to make a point is a cardinal sin, but even inadvertently giving the audience a false impression by inaccurate or sloppy presentation of data or a failure to adhere to conventional use of words, tables, and figures is as bad as guessing at measurements in the laboratory. Precision, accuracy, and honesty must continue throughout data analysis and presentation.

The ease with which data can be analyzed by computer and the comparable ease with which you can devise a table or a graph with the computer are invaluable advancements for scientific communications. However, any technology involves both beauty and danger: Harnessing electricity to provide light or control temperature in a room is a beautiful thing, but misused, the same electricity can injure or kill. The computer should not dictate the kind of analysis and presentation you will use. Recognize the conventions used in tables and illustrations of all kinds. The readers will interpret data according to conventions they are familiar with or by their visual perception of what appears most important. Tufte (1983) describes some of the pitfalls that occur with presentation of data in figures. These misconceptions result from visual or psychological illusions because of the arrangement of data relative to simple communication devices involving color, intensity, size, and spacing. Any amount of explanation will not completely overcome the psychological impact of a wide line compared with a narrow one or a large letter compared with a small one. Size suggests importance. The same would be true of a column of information in a table if it were set apart in some way with color or spacing. Guard against misleading an audience with poorly presented data. With and without statistical analyses, your data can be complex. Your job is to simplify your message without falsifying the data. Every scientific experiment has exclusive variables or details that need to be emphasized, and results may require unique methods of presenting data.

Because of the many differences in portrayal of data, you should study the journals published in your scientific discipline. Follow any suggestions made in your style sheet. With electronic tools for your own use and with publication by electronic means, it is critical that you check the most up-to-date guidelines if you plan to publish. These guidelines or the editor can tell you what is acceptable and how to submit the form of data that you use. Many of the requirements typical just a few years ago are now obsolete. As the Science Illustration Committee (CBE, 1988) concludes: "The simplest rule of thumb for the author and illustrator should be: follow what you see in print." But be sure the printed versions you are looking at are recent ones. Note examples of tables and figures in Appendix 10. The following observations may also be helpful.

TABLES

If a table is used, don't repeat exactly what it says in the text—just call attention to its main points. A table should be able to stand alone, that is, communicate a point or points without the need to refer to the text. Background material such as information on how an experiment was conducted will be given in the text, but the reader should be able to interpret data presented in a table without referring to the text. Abbreviations that cannot be used in the text are sometimes appropriate for tables, but meanings that are not immediately clear should be defined in the caption, a headnote, or a footnote. For clarity and understanding, a table may repeat information in more than one form; that is, in addition to the absolute values presented, it may give percentages, totals, means, averages, or ratios of those values.

Authors are more likely to put too much rather than too little information in a table. Limit the items in the field, especially in tables used for slide presentations. Often full columns of information can be omitted. Be suspicious of any column that has few or no data points that differ. A string of zeros or a column that repeatedly notes the same result can often be omitted. A full table consisting of no more than six or eight data points can usually be replaced by a sentence or two in the written text; however, the same table may serve as a good visual aid in a slide or poster presentation. Tables are also overloaded when headings and footnotes are too heavy. For example, if an abbreviation is obvious from information in the caption, that abbreviation can be used with no further explanation.

Problems that occur with tables may be crucial to a reader's understanding. Errors can occur easily in transferring data to a table or in subsequent revisions. Captions, similar to titles, are sometimes too long and wordy; they should be concise and contain only the key words needed to clarify the message in the field of data. Numbers should show no more decimal places than are essential for reasonable precision and accuracy. As with other communication, try to look at your tables from the viewer's perspective. Make them precise, concise, simple, and clear. The following briefly illustrates parts of a table and the terminology used to name those parts:

Table No.	Caption or title		
	Main boxhead (identify items in field)[a]		
	Secondary head No.1[b]		Secondary head No.2
Stub Head	Tertiary No. 1	Tertiary No. 2	
Stub #1[c]	Field item No. 1	Field item No. 2	Field item No. 3[d]
Stub #2	Field item No. 4	...	...
...	...	...	...

[a,b,c,d]Footnotes in order from top to bottom and horizontally.

Characteristically, a table contains no vertical lines. Many computer programs provide an option to use grids in which to plot data. Look at the journals published in your discipline. Unless they regularly publish tables on grids, don't choose this option. Grid lines are ink used for no real purpose. Three horizontal lines run the full width of the table, one beneath the caption and any headnotes, one beneath the headings for the stub and the field, and the third below the field and before any footnotes. Other horizontal lines, called straddle rules, run across all the columns of items to which the heading above the straddle line refers.

The boxheadings identify items in columns, which should be the dependent variables, and the stubhead identifies the independent variables for items in horizontal rows. Comparisons between like elements in the data should be made down columns, not across rows. Generally, the main boxhead should define the meaning of the items in the field by telling us whether those items are percentages, yields, concentrations, colors, or other measures. Any subordinate boxhead simply adds precision to the main head. Use as few headings as possible to communicate clearly; only rarely should you go beyond tertiary headings and not beyond secondary heads if possible. Be sure to include units of measure. For example, the main heading might read "Rate (mg L^1)"; numbers in the field would be rather meaningless without the mg L^1.

The sequence of footnotes is carried from top to bottom horizontally across each line of entries in the table. Traditionally, for publication, tables are typed on pages separate from the text, and a marginal note is made beside the first mention of the table in the text so that it can be positioned appropriately in the layout for the printed manuscript. Check journal guidelines to be sure this technique has not changed for electronic submission.

Excellent information on tables is presented by Day (1998), the *CBE Style Manual* (1983), and Smith *et al.* (1980). See sample tables in Appendix 10. For the specific stylistic details for your discipline, consult your own journals and style manuals. Unless your style sheet specifies otherwise, you can use the following guidelines:

Preparing Tables

For Publication

1. First study the publisher's style sheet carefully and look at examples in the text.
2. Type each table on a separate page with double spacing throughout.
3. Use arabic numerals to number the tables.
4. Prepare tables with the similar items, or items you want to compare, reading down columns, not across lines.
5. Group items logically, with control values to establish a base line for comparisons. Any gradation or trend in data should probably be stressed.
6. Round off numbers, and don't use excessive decimal places. Decimals must be aligned in columns.
7. Be sure the caption is descriptive of the table's contents. No verb is necessary in this caption.
8. Explain in the caption, headnotes, or footnotes all of the nonstandard abbreviations and symbols used.

9. Verify all information. Make sure that all data and statistical analyses are accurate. Verify again after any revision or transfer of information.
10. Check tables for accuracy in use of symbols, units of measure, and other labeling. Be consistent with such labeling among all tables and figures in the same text.
11. Proofread carefully.

For Slides and Posters

Most of these suggestions apply equally to tables that are used in written reports or for other presentations. Simplicity is even more essential because the viewer of a poster or slide presentation is not going to be able to study the table as long as the reader of a manuscript can. In publication, the use of symbols, shapes, sizes, and especially color may be limited, but the slide or poster can make important use of such symbolic language. If you are selective in the media you use and if you present your most representative data, your tables will lead the reader or viewer to the conclusion reflected by the full data. That same remark could be made about figures.

FIGURES

Illustrations, or figures, come in various sizes and shapes. Photographs, drawings, flow charts, line graphs, bar graphs, pie charts, maps, and variations of these forms can add an understanding that is difficult to convey in words. In considering where and how to use an illustration, keep in mind the values and standards we are striving to achieve in scientific communication. Consider the purpose and what form will best fit that purpose. Look at the nature of the data—numbers, ranges, differences, appearances—and what point you should make with the data. Try to determine how your audience will respond to the illustration. Will they see something different from what you see because they are not familiar with the information? Test your portrayal of data on someone not familiar with the point you wish to make. Be familiar with the conventions used with illustrations, including symbolic meanings that are expressed with lines, sizes, bars, numbers, colors, shading, and axes, as well as words. And understand any constraints in producing clear communication, whether they be the possible reproduction of a photograph, the capabilities of you or a graphic artist, or the limitations of the equipment your publisher or you use. With all these things in mind and with an honest and sincere effort, you can produce an illustration that carries a simple, accurate message.

If illustrations are not clear, accurate, and appropriate, they are mere distractions and should be avoided. If you need to display specific values, a table is usually the better choice. A great deal that has been written about charts has to do with problems involved in their use, whether it be the difficulty in presenting the data, the difficulty in interpreting the data, or what Tufte (1983) calls "chartjunk" or others might call "noise." Anything that distracts, anything that is simply decorative, anything that is illegible, or anything without a positive communication value should not be used with scientific illustrations any more than such should be used with words or the data themselves. Be sure that the end result of your illustration is that the audience concentrates on what is communicated, not how the image looks.

Computer technology has made possible quick and accurate creation of figures without the graphic artist. Find a graphics software program that works well for your kind of data, and then use it carefully. Most programs will allow you to draw boxes, shadows, and three-dimensional bars. The programs also provide some atrocious designs to fill the bars or place them in multiple rows with depth perception. Keep your graphs simple; data are often difficult enough to comprehend without challenging readers to see bars hidden behind bars or distracting them by strange and indistinct designs. **Data are best presented in the simplest possible form.** Check samples in the journals in your discipline to determine forms that your editors prefer.

Designing illustrations and the graphic presentation of statistical data are subjects for entire books, and all of us could well benefit from a course on those subjects. The most that I can do is to introduce you to standards and values that I have encountered in working with scientific communications and to suggest other sources of information. First, by all means, check *Illustrating Science* (CBE, 1988) and Briscoe (1996). For an introduction to simple graphs, MacGregor (1979) is still good, easy to read, and well illustrated. Tufte (1983) has produced an interesting study of visual display of data. His book includes some history of the art and discussion of problems such as the inadvertent "lie factor" and chartjunk that cloud graphic communication. He illustrates his points well with examples from published charts. Schmid (1983) also presents concerns about the problems in graphic accuracy and in facility of communication. He uses good illustrations and covers the material on graphs and maps well. In the biological and medical sciences, illustrations, drawings, and photographs of life forms are often depicted. Hodges (1989) provides good material on such illustrations.

Explore these sources, and you will come away with a basic understanding and respect for standards in visual display. After you have looked into at least two or three of these, look more closely at the illustrations in the journals in your area. You will find errors there, but you will also find what works best for the presentation of your data. Along with the information from CBE (1988) and Briscoe (1996), consider the following basic principles for producing graphs and other figures, and note the examples in Appendix 10. Remember that modifications of these principles may be needed if your data demand a different form. However, always choose the simplest form possible to make your point clear to any audience.

Preparing Graphs and Other Figures

1. Be sure the graph carries your point better than the text or a table would. If the figure becomes complex and requires extensive explanation, reconsider, divide the data, or try something different.
2. Consider what kind of figure you need. Do you need a photo, a line drawing, or a graph? What kind of graph would be best—line, bar, or pie?
3. Make it simple. A figure should be comprehended at a glance. Draw graphs to agree exactly with experimental data, but don't overload them with information.
4. Limit the number of curves or bars on a graph. A single figure can hardly satisfactorily communicate clearly with more than three to five lines or eight to 10 bars (more than 10 bars can be clear if they are grouped).
5. Plot any independent variable on the horizontal (X) axis, or abscissa, and the dependent variable on the vertical (Y) axis, or ordinate.
6. Avoid wasted space. Scale details to agree with the data, but do not extend the axes beyond the point needed. Put the legend into the field of the graph if possible, and don't put a box around it.
7. Label all axes carefully and show units of measure. Use tics and subtics to subdivide the axis so that it is not overcrowded with numbers.
8. On bar charts and linear line graphs, most scales should start with zero. If you must compress scales with slash marks or start beyond zero, be sure this modification is clear.
9. Remember that position, size, shape, length, symbols, angle, and color are all visual codes that carry messages to a reader. Do not let them convey the wrong message.
10. Select the size and format to fit the journal or other use for which the visual is intended. (Avoid color for most publications; use color in posters and slides.)
11. If a set of graphs is used in the same paper, poster, or slide set, be consistent and uniform in your use of all visual and verbal codes.
12. For a journal publication, consult your editor or follow guidelines for authors. Electronic publication may determine how your figures, as well as tables and text, should be submitted.

Bar Charts

1. Bar charts often have just one measurable axis. If they have two, they are sometimes called histograms.

2. Bar charts can be presented for data collected at even or uneven intervals.

3. The bars should be wider than the spaces between them.

4. Use conservative patterns (solids, shading, or hatch marks) to differentiate bars. For posters and slides, color may be preferable.

5. Show significant differences with a least-significant-difference bar or with letters or asterisks above bars.

Line Graphs

1. Line graphs should have two axes. Avoid a third axis if at all possible.

2. Simple line graphs should present data collected at regular intervals to show trends without extrapolation between data points.

3. The curves themselves should be the boldest lines. Axes and tic marks should be less bold.

4. Be careful with line patterns. Dots, hyphens, or slashes can become confusing. Sometimes it is better to use distinct symbols (●, ■, ▲) rather than line patterns. Use distinctive colors for lines on posters or in slides.

5. Plot the length of intervals on both axes so that slopes are not excessively flat or steep.

Although bar charts and line graphs are probably the most common forms for figures in scientific journals, other illustrations are also useful. Maps, flow charts, diagrams, line drawings, chromatographs, photographs, and other forms can sometimes communicate much more than the text alone. *Illustrating Science* (CBE, 1988) is a good source of information on all these illustrations, and Imhof (1982) is a master of cartography. Both Briscoe (1996) and Hodges (1989) present very good information on techniques for creating illustrations, especially in the biological sciences. Whatever illustration you use, be sure to submit high quality copies for publication and for use in slides and on posters. An out-of-focus photograph or a graph filled with chartjunk (Tufte, 1983) will merely distract from what you wish to say. You may also need assistance from a graphic artist for line drawings or other figures, but be sure you understand enough about the illustration yourself that you will know what it is communicating. Symbolic communication with or without accompanying words is a strong form of communication. Be certain that it carries the right message.

References

Briscoe, M. H. (1996).*Preparing Scientific Illustrations: A Guide to Better Posters, Presentations, and Publications*, 2nd ed. Springer, New York.

Council of Biology Editors (CBE) (1983). *CBE Style Manual*, 5th ed. CBE, Bethesda, MD.

CBE. (1988). *Illustrating Science: Standards for Publication*. CBE, Bethesda, MD.

Day, R. A. (1998). *How to Write and Publish a Scientific Paper*, 5th ed. Oryx Press, Phoenix, AZ.

Gastel, B. (1983). *Presenting Science to the Public*. ISI Press, Philadelphia.

Hodges, E. R. S. (1989). *The Guild Handbook of Scientific Illustration*. Van Nostrand Reinhold, New York.

Imhof, E. (1982). *Cartographic Relief Presentations*. Walter de Gruyter, New York.

MacGregor, A. J. (1979). *Graphics Simplified*. University of Toronto Press, Toronto.

Schmid, C. F. (1983). *Statistical Graphics: Design Principles and Practices*. John Wiley & Sons, New York.

Smith, R. C., Reid, W. M., and Luchsinger, A. E. (1980). *Smith's Guide to the Literature of the Life Sciences*, 9th ed. Burgess, Minneapolis, MN.

Tufte, E. R. (1983). *The Visual Display of Quantitative Information*. Graphics Press, Cheshire, CT.

12

ETHICAL AND LEGAL
ISSUES

"Whatever the rationalization is, in the last analysis
one can no more be a little bit dishonest than
one can be a little bit pregnant."

C. IAN JACKSON

Professional ethics have to do with one's professional behavior as it affects others. Ethics are determined by cultural, social, and professional values and may not be governed by laws. Because human values dictate ethics, standards and beliefs can differ among individuals, groups, and cultures. Professional behavior relative to communication may be governed by policies or codes of ethics of a society, a company, an agency, or a publisher of your work, and your ethical decisions may sometimes be guided by those policies. But the final responsibility for your actions rests with you. Laws supposedly apply the same for all those ruled by the same governments, yet both laws and ethics result in behaviors important to human interactions. In scientific communication, professional ethics, copyrights, and patents are the main issues of ethical and legal concern.

ETHICS IN SCIENTIFIC COMMUNICATION

Most scientists maintain high ethical and professional standards. As scientists, they are interested in discovering truths, and any communications among them

125

must be carried out with a high degree of accuracy and integrity. It is in their best interests to protect the pool of scientific knowledge from any fabricated data or conclusions based upon insufficient evidence. But human errors do occur, some people do not value honesty, and good things can happen to bad people. Isolated instances of scientific fraud have been found even among previously very reputable researchers. Every scientist is responsible for protecting the integrity of science.

In scientific communication, two kinds of ethical errors are unforgivable—distorting your own data and plagiarizing the work of others. Other breaches of integrity are also deplorable but sometimes hard to diagnose. Most difficult is recognizing your own bias and admitting that even you can rationalize truths. It is not always easy to mark the dividing line between truth and assumption, and all of us can err. Such errors can be made inadvertently with no intent to lie, but for that reason they are all the more important and all the less forgivable. You have to discipline yourself to demand care, accuracy, and objectivity in every instance.

Distorting data intentionally is inexcusable, but that which is not intentional can also be detrimental to science and to your reputation. It's easy to see the dishonesty when you change numbers in a data column, but what of the situation in which you believe a result should occur, the data almost point in that direction, and deleting one experiment would take care of that *almost*. Can you delete that experiment? Was that odd data set simply an anomaly that should not be considered anyway? Is there a typographical error or erroneous transcription in your data? Do you have time to experiment further to substantiate what you believe should have resulted? These questions are not easy to answer, and often no one except you **can** answer them. You can become so overly meticulous that you are ineffective, or you can rationalize your way into really dishonest reporting. Your scientific integrity is based upon careful scrutiny of your work and sound judgment and objectivity in how you answer the questions involved.

The Committee on Science, Engineering, and Public Policy (1995) suggests that "social and personal beliefs—including philosophical, thematic, religious, cultural, political, and economic beliefs—can shape scientific judgment in fundamental ways." Individual and cultural prejudices have troubled scientific research and reporting for centuries, and we are not without prejudices today. In *The Mismeasure of Man*, Gould (1981) exposes the social prejudice that has produced "documented" studies on biological determinism to prove that some groups of people are inferior to others (women to men, for example). As "proof" these studies have used selected criteria that fail to reflect a total physical reality. Gould's (1981) conclusions should bring a degree of humility to all of us: (1) "no set of factors has any claim to exclusive concordance with the real world," and (2) "any single set of factors can be interpreted in a variety of ways." Think about those remarks when you conclude that you have discovered "truth."

Unintentional dishonesty can also affect communication. We know when we speak or write an outright lie, but we must also be sure our statements are not misleading and ambiguous. A major difference between scientific writing and creative writing is that in creative writing we can allow the readers to interpret as they choose. Ambiguities and double meanings simply add interest. We are not allowed that avenue in scientific communication. Read your sentences carefully to be sure that they will not be easily misinterpreted. The same is true for tabular or graphic portrayal of data. An inaccurate number or a visual deception in a graph can misdirect a reader. Your responsibility is to know the conventions for data presentation and check the accuracy of all details. Careful research, use of scientific reasoning, an open mind, clear and accurate communication, and a willingness to be honest at all costs will generally result in good ethical conduct. Settle for nothing less in yourself and your colleagues.

Plagiarism is a dirty word in scientific writing and is as inexcusable as is distorting data. It is both a legal and an ethical issue. A dictionary defines the term as some kind of literary theft or stealing. Ethically, it is disrespect not to recognize the ownership of the property of other writers. Like other forms of dishonesty, plagiarism is easy to distinguish when it is blatant. You lift words from someone else's work without giving due credit. That is clearly plagiarism. But how about ideas or words that are not quite the same but mean the same thing, or how about picking up from your subconscious something you have read, thinking it is your own idea? These nebulous situations occur, and although we cannot always be afraid that we may be using ideas that are not entirely original, we must give due credit for both words and ideas. To guard against plagiarism or a failure to document sources, be familiar with the literature in your discipline.

The Internet has good information to help you understand what constitutes plagiarism. I recommend that you look first at the site of Gordon *et al.* (2003) at http://www.zoology.ubc.ca/bpg/plagiarism.htm#sel. More briefly, Dartmouth College's Committee on Sources (1988) has outlined three types of plagiarism. The committee lists not only "direct quotation or word for word transcription" but also "mosaic or mixing paraphrase and unacknowledged quotation" and "paraphrase and/or use of ideas." These three forms are all bad, but it may be that some students and professional writers do not recognize the last one as plagiarism. Of course, you should refer to the opinions or results from other researchers and give due credit, but too much reliance on another's ideas to establish your own paper or just reporting another writer's ideas into your own words with or without citation and documentation can constitute plagiarism as surely as if you had used their words.

Plagiarism can also result from simple ignorance in how to reference other works or in sloppy, careless documentation. Careful documentation and verification of your references as well as clear knowledge of what has been done in your area of study are essential. Learn the accepted conventions for documenting the work of others and for obtaining copyright permission. When you question your right to use the words or ideas of others, find answers to those questions before you proceed. Some answers may be found in this text, but when questions remain, go to other sources. Ignorance is no excuse. I suggest international students whose cultures may have different ethical codes for plagiarism read Chapter 20 and study the standards to use while you are in the United States.

PROFESSIONAL RESPECT FOR OTHERS

Other questions of professional ethics have to do with your relationship with other scientists and with any professional group to which you belong. Candid sharing of scientific findings and maintaining the confidence of others are essential to your being an ethical scientist. The society or registry in your discipline probably has a code of ethics that you can study and use as a model. Be conscious of what constitutes fraud, conflicts of interest, and the need for confidentiality. Your colleagues soon learn to what extent you can be trusted.

Giving due credit to others is important in your reference to their work and in recognizing colleagues as coauthors. It is equally important that undue credit not be given. Questions of junior or senior authorship and of who should be included as an author are not always easy to answer. If you are a graduate student or a junior scientist, discuss with your advisor the authorship and order of authors on a paper or presentation. The *CBE Style Manual* (CBE, 1983) says, "The basic requirement for authorship is that an author should be able to take public responsibility for the content of the paper." That means that any coauthor should be clearly involved in the research and composition and be able to answer questions asked about methods, results, or any other idea expressed in the paper. Despite what some may call honorary collaboration or honorary authorship, any author listed on a paper should have contributed substantially both to the work and to development of the manuscript or presentation. Giving false authorship a nice name does not make it any more ethical. For information on authorship and making ethical judgments, read Macrina (2000) or *On Being a Scientist* (Committee on Science, Engineering, and Public Policy, 1995).

Ethics are matters of making good decisions about questions of appropriate conduct regarding your work and that of others. There is often no legal retribution for offenders of codes of ethics, but your ethics relative to your behavior in your association with colleagues are instrumental to your reputation and the acceptance of your communication efforts. Your colleagues and your professional society may provide some guidelines, but unfortunately, no definitive rules regarding ethical judgments exist. Almost every situation demands an individual consideration, and you must draw your own conclusions. Consider the following issues that you will probably encounter in your career as a scientist.

Respect Your Data

Scientific progress is a continually evolving process. Your work today depends upon results of work done in the past, and the future will also look to work done today. Scientists must be able to trust the honesty of their predecessors and colleagues in reporting data. Don't make up your mind about your results before your data are collected and analyzed. Trimming, cooking, or otherwise manipulating data is as bad as reporting false data (Fig. 12-1). Be careful with your own prejudices, and be cautious when you work with others, especially with specialized scientists. In working across disciplines, researchers have to trust each other for ideas and analyses that their own disciplines do not include. The same is true with specialized statisticians who often know little about your science but can plot and analyze your data in a variety of ways. They can contribute invaluable expertise to a project, but they may not recognize the relative importance

FIGURE 12-1
Trimming or cooking your data is as bad as reporting false data.

of your scientific variables. You need to understand links between your work and that of your colleagues and to be familiar with the meaning of any statistical analysis used with your data.

Be Careful with Confidentiality

In your own laboratory, information may be available that should not be disclosed to others until experimentation is complete and researchers are ready to publish their findings. Or you may review a paper for someone and find interesting information that should be held in confidence until that paper is accepted and published. Also, a pending patent may make it unwise to disclose information. Scientists have an obligation to share information, and the obligation to keep information in confidence for a funding agency may result in a conflict of interest. All parties involved need to understand when disclosure of information is acceptable. At professional meetings, scientists should certainly exchange ideas, but take care in making positive pronouncements on unsubstantiated results. Disclosure before publication should be handled with caution.

Don't Publish the Same Thing Twice

To publish the same data in more than one scientific outlet is not only unwise but contrary to most publishers' policies. If you need to convey the same information to two audiences, maybe the scientific community and the public, two publications might be based upon the same experimentation but presented in different ways, and it may be important to report to the publishers that the information appears elsewhere. To submit results of the same research in two scientific journals is unethical even if you completely rewrite the text. Too much scientific literature already exists to ask an audience to read the same thing twice, and journal editors and reviewers put too much time and effort into getting a paper ready for publication to be told you are having it published elsewhere. Be sure one publisher has released your manuscript before you ask another to consider it.

Acknowledge Your Errors

You are human; you can make mistakes. If, in all sincerity, you say or write something that you find later is misleading or untrue, try to right the situation by acknowledging that error. To make a mistake is one thing; to allow people to continue to be deceived is another. Your colleagues are also human; they make mistakes. And they will respect you far more if you acknowledge an error than if you let it go, even if the intent to deceive was never present.

Support an Ethical Work Place

When you agree to work for a company, an agency, or an institution, you are also agreeing to abide by the policies of that employer. Many of them have specific requirements relative to communication of information. You may need to have approval for any paper published, speech made, or meetings attended. The employer may also trust you with confidential information that you should protect from unwise disclosure. Sometimes your own values may not concur with the policies of your employer. Ethical decisions on what to do can be difficult. Know your employer's policies and abide by them insofar as they are within the law and conform to your ethical and professional standards. If you are uncomfortable with your employer's policies, you would do well to find another job.

Respect the Time of Others

If you are asked to contribute to a paper as a coauthor, get your part done expeditiously so that you do not delay the publication for other authors. The same is true when you review a paper; get the review completed in a timely manner so that the author or editor is not unduly delayed. You should also be considerate of a fellow scientist's personal and professional time. Lengthy consultation should be a matter of mutual agreement and benefit. If someone generously provides time for consulting about your research, be sure that person is not coerced and that you acknowledge contributions made.

Watch Out for Conflicts of Interest

Like plagiarism, blatant conflicts of interest are easily recognized. Trying to mix a personal relationship with a professional one or overlapping two professional activities can result in conflicts of interest. A personal or business relationship should not interfere with your work as a scientist and a professional. Business partnerships with an advisor who has some authority over your degree program or research are questionable, as is having close relatives serve on your graduate committee. Remember that there are no definitive rules: A father and daughter could become a highly ethical research team. But treat any possible conflict of interest with great caution. The scientist must avoid any personal bias in his or her judgments.

Be Fair with Your Time and Effort

Conflicts of interest can involve personal and professional relationships with others, or you may find issues in your individual work competing for your time, energy, and effort. Scientists can become involved in many activities—research, teaching, grant seeking, professional obligations to society, speaking engagements, seminars, workshops, travel, and myriad other activities. Quality of any one activity can suffer at the expense of the others. To assume a responsibility and do a poor job can be worse than refusing to assume the responsibility. Only you (and perhaps your boss) can determine where your responsibilities lie and how far you can extend your time and energy.

Avoid the Sin of Omission

Refusal to assume a reasonable work load can be equally as irresponsible as trying to do too much. It may be hard to believe that you can be accused of unethical actions if you do nothing, but breaches of conduct come with omission as well as commission. As a scientist, you will often be asked to contribute your expertise to a collaborative project. You may be asked to serve on professional committees or to be a reviewer or associate editor for a journal. You will have to be discreet in determining how much you can do and still maintain quality work. You cannot yield to every request for your time and energy; on the other hand, to reject all requests to serve your science or your profession is just as unethical.

Watch the Company You Keep

You may be "Honest Abe," but if you collaborate in your research with unscrupulous individuals, their reputations will rub off onto yours. Be alert to how your colleagues are received in the scientific community. Also, as you work with a fellow scientist, observe the care he or she takes with accuracy and the truth. You can usually recognize unethical behavior if you are conscientious.

Be Firm with Your Own Ethical Standards

If you have not had an opportunity to take a course in ethics, consult a text such as those of Bayles (1989) and Macrina (2000), and be sure you can define ethics relative to your values and those of your profession. Professional reputations and scientific values depend upon scientific integrity. Objectivity and scientific methods are only as honest and unbiased as the researcher or communicator who uses them.

Beyond a reasonable point in the critical review processes, we have to trust each other to be honest and accurate. Generally speaking, scientists are honest people. Those who cannot be trusted in their experimentation or reporting have no place in the world of science. But devils do exist, and we must constantly be alert to unethical and unprofessional behavior. The least you must do is to control your own behavior and uphold the highest ethical standards. As Thomas Carlyle said, "Make yourself an honest man, and then you may be sure that there is one rascal less in the world."

THE LEGAL ISSUES: COPYRIGHTS AND PATENTS

Sometimes the message you want to communicate to other scientists is a simple "it's mine," or you may wish to make use of a creation that belongs to someone else. In either situation, you are dealing with legal, as well as ethical, issues. Laws governing copyrights, patents, and trademarks protect the property of individuals and groups. Scientists must use each other's ideas and inventions or we wouldn't make much progress, but be sure you give credit to your sources and ask permission when you make extensive use of the work of others. You will also be asked to grant permission to others. Know your legal rights, but recognize the need to exchange information with other scientists. A few notes on copyrights and patents may be helpful.

Copyright

You own the copyright on any **tangible expression** that you create. Along with other forms, tangible expressions include written words, illustrations, printed works, electronic software, and recordings. Copyright does not cover the ideas, procedures, processes, concepts, or discoveries contained in such works. Protection of those things can sometimes be obtained through patents. Copyright registration is not a condition required for copyright ownership. Statutory copyright begins when the document is created, but protection and appropriate use of your work are increased when you have the copyright registered. General recommendations are to have your copyright registered within 5 years of the creation of a work and to avoid any law suit if the creation is not registered.

If a work is not yours, find out whose it is before you use it. The author involved with scientific communications needs to be conscious of copyright privileges and regulations. You will need to know how to register copyright for yourself, how to grant permission, and how to obtain permission for use of works from other copyright holders. Below are some handy things to know, but consult the copyright law or a lawyer when you have important questions on copyright.

—**Both published and unpublished works** in any tangible medium of expression are protected by copyright and may be registered with the U.S. Copyright Office. Whether it is registered or not, copyright protects only the expression of an idea (the words or other tangible expression), not the idea itself.

—**Registration of copyright** can be obtained any time throughout the copyright duration. For information on registration of copyright, you may write to the U.S. Copyright Office, Library of Congress, Washington, DC 20559. Or check the Web site at www.loc.gov/copyright/. For written scientific work, you will want the forms under nondramatic literary works.

—**Copyright lasts for a certain duration:** Works put into tangible expression on or after January 1, 1978, are under copyright until the death of the author plus 70 years thereafter.

—**Works for hire** are those expressions or inventions that an employer requires or commissions you to create as a part of your job. In other words, you are being paid to create something, and any copyright, patent, or other benefit that comes from the creation belongs to the employer.

—**All works prepared by officers or employees of the U.S. government** as part of their official duties are in the public domain. You don't have to obtain permission to use these materials, but you should still document your source.

—**"Fair Use"** (although not clearly defined) generally means use for your own educational purposes as opposed to use for commercial gain. This definition is certainly oversimplified and incomplete. If you question your rights to use the works of others, consult an authority or apply to the copyright holder for permission to use the work.

—**Electronic communication** has added a new dimension to copyright. Software bought by one user should not be copied by another any more than a book should be copied for a friend or colleague. New laws are being considered for how to copyright and document electronic communication. Be alert to developments in this area.

To grant copyright permission. From offices of journals in which your work is published, you may receive a form that you and any coauthors will sign to transfer copyright or to grant printing and reprinting rights to the journal. Some publishers may ask that you transfer the copyright itself so that they are the owners, and you must refer any request for permission to use the work to them. Others are simply asking that you grant them permission for reprinting and distributing the work. In this case, you still own the copyright.

If individuals ask for permission to use your work or a part of your work, they may send a form to be signed or a letter requesting permission. Be sure the exact portions of works to be used are fully described or copied and that the description or copy is kept with the signed permission granted. Keep this information on file along with specific details on where and when the material will be used. Request that persons using your material clearly acknowledge both the source from which it came as well as you and any coauthors.

To obtain copyright permission. When you know who the owner of the copyright is, write directly to that person, publisher, or agency for permission. You should begin efforts to obtain any copyright permission as soon as you know that you want to use the material. You may want to make a phone call to locate the copyright holder and to see whether he or she is amenable to your using the material. However, a phone call will **not** constitute evidence that you have been granted the permission. You need specific information in writing, and this process may take weeks or months. You can delay your own publication by waiting too late to apply for copyright permission.

In requesting permission to use another author's or publisher's work, clearly identify the specific material to be used, including:

1. Author(s)
2. Title
3. Date of publication
4. Publisher
5. Specific selection to be used
    a) The form (as table, photograph, text)
    b) A description of the content (perhaps include a photocopy of the material you will use)
    c) Pages on which it appears in the original text

Tell the agency or person to whom you apply where, when, and how you will use the material requested; whether your work will be published and what form it will take; and who the audience will be. Indicate that you will give full credit to the author(s) and publisher, and be sure that you do so. You will often use a form letter in obtaining copyright permission and will want the information in duplicate for your files as well as the grantor's. Ask for signatures of all authors. Keep a clear record of permission received. A sample form letter or other information on copyrights is found in the *CBE Style Manual* (1983) and in Dodd (1997). A sample letter requesting copyright permission is in Appendix 11.

Patents

Patents are more versatile and offer more protection than do copyrights where they are applicable, but they are not applicable to the expression of ideas. Patents protect the ideas as they are put into practice as machines, manufacture, processes, or composition. Copyright protection begins as soon as the expression is created; patents must be registered to serve as protection. Copyrights protect for longer periods than the 17-year protection for patents. Work for hire is applicable to both copyrights and patents. The remarks here on patents apply just to U.S. patents.

For the sake of simplicity, I refer to any patentable creation as an "invention," but patents can be obtained on diverse items, including synthesized materials and life forms. As with copyright for electronic communications, biotechnology has added a new dimension to patents. It is now possible to patent life forms if they have been synthesized by human efforts and do not exist in nature without the human intervention.

Any invention that is patented must be proven to be **novel, not obvious, and useful.** In applying for the patent, you must give evidence that your invention has all three of these characteristics. **Novel** simply means that your invention is new, that a like invention did not exist before yours. **Not obvious** means that what you have put together is not something that anyone might immediately derive from the same materials. Your invention is needed, but no one else has come up with a way to fill that need. Finally, in filling this need, you have made something that is **useful.**

Several communication efforts are important relative to patents. First, you must disclose to the U.S. Patent and Trademark Office any information you have on how the invention works, its components, and any similar inventions already in existence. That address is Commissioner for Patents; P.O. Box 1450; Alexandria, VA 22313-1450. These disclosures require that you carefully search and read the literature on your subject. Any failure to disclose similar inventions already in existence can thwart your chances of obtaining a patent even if you did not know about the existing invention.

An important question relative to communications and patents is when to publish information about your invention. If you describe your invention to the public before you have registered the patent, it is considered public information and will probably not be patented. Be careful about presenting a paper on your invention at a professional meeting, publishing a journal article about it, or even describing it to anyone beyond confidential disclosure to colleagues who can be trusted to respect your claim to ownership. Although the same is not true in many other countries, in the United States if you publish information about your invention before you realize that you should obtain a patent on it, you can apply for a 1-year grace period for filing the patent application. Remember that you must be able to prove that you invented it first; someone else may already be using the idea or even applying for patent. Records that you keep as you are working on the invention can be essential to proving that the invention is yours. Keep careful, dated records, and if you foresee that you might be seeking a patent, have witnesses sign and date information that you record along the way.

Finally, perhaps most important to scientific communications is the patent literature. The disclosures on how inventions are produced and how they function are published and offer a storehouse of information helpful to other scientists. Paul (1975) claims that "The U. S. patent literature is the largest and most comprehensive collection of technical information in the world." Certainly, the information on patents contains useful ideas, and you should not neglect to explore your research subject in this area.

If you are thinking of patenting an invention, you are sure to have numerous questions that I cannot answer. You may obtain some answers by going to the government Web site, www.uspto.gov. Complexities arise in the patenting process that often require the services of a lawyer specialized in the area. In fact, one of the first things you may want to consider is hiring a patent attorney to deal with the patent office in proving that your invention is novel, not obvious, and useful. Books on the subject of patents that may also prove useful include that of Wherry (1995), with basic definitions and information on patents as well as trademarks and suggestions for searching patent literature. Auger (1992) deals with both national and international issues on patents, and Levy (1991) provides his ideas on what to do with your own ideas. Chapter 9 by Thomas D. Mays in Macrina (2000) offers succinct information on both copyrights and patents as well as other ownership of intellectual property.

References

Auger, P. (1992). *Information Sources in Patents*. Bowker-Saur, London.

Bayles, M. D. (1989). *Professional Ethics*, 2nd ed. Wadsworth, Belmont, CA.

Committee on Science, Engineering, and Public Policy (1995). *On Being a Scientist*. National Academy Press, Washington, DC.

Committee on Sources (1988). *Sources: Their Use and Acknowledgment*. Dartmouth College, Hanover, NH.

Council of Biology Editors (CBE) (1983). *CBE Style Manual*, 5th ed. CBE, Bethesda, MD.

Dodd, J. S., ed. (1997). *The ACS Style Guide: A Manual for Authors and Editors*. American Chemical Society, Washington, DC.

Gordon, C.H., Simmons, P., Wynn, G., and the Arts Faculty, University of British Columbia (2003). "Biology Program Guide 2003/2004: Plagiarism." www.zoology.ubc.ca/bpg/plagiarism.htm#sel (verified July 2, 2003).

Gould, S. J. (1981). *The Mismeasure of Man*. W. W. Norton & Co., New York.

Levy, R. C. (1991). *The Inventor's Desktop Companion: A Guide to Successfully Marketing and Protecting Your Ideas*. Invisible Ink Press, New York.

Macrina, F.L., ed. (2000). *Scientific Integrity: An Introductory Text with Cases*, 2nd ed. ASM Press, Washington, DC.

Paul, J. K. (1975). *Fruit and Vegetable Juice Processing*. Noyes Data Corp., Park Ridge, NJ.

Wherry, T. L. (1995). *Patent Searching for Librarians and Inventors*. American Library Association, Chicago.

13

SCIENTIFIC
PRESENTATIONS

"Nothing clarifies ideas in one's mind so much as
explaining them to other people."

VERNON BOOTH

Presentations at professional meetings and elsewhere are extremely important
to scientific communication and to your individual reputation. Scientists build
on the discoveries of others through communication, and your own speaking
ability may mean the difference in your getting a job or a promotion. Often,
many applicants for a position will have equally impressive academic knowledge and technical skills; the ones who get the jobs are those who also present
themselves well. Take advantage of graduate school as a time to perfect your
speaking skills. You will probably be required to make one or more presentations
while you are in graduate school. You may even have a course in your department in which you and other graduate students regularly deliver presentations.
Also, attend seminars and special lectures in other departments and observe
good and bad techniques in delivery as well as the content of the speech, and
use every opportunity available to take presentations to professional meetings.
Valuable experience can be gained by attending presentations either as a speaker
or a listener.

If you have occasion to attend regional or national meetings sponsored by
societies allied with your discipline, go and present posters or slide talks. In doing
so, you are establishing your reputation with colleagues, and you may encounter

prospective employers for whom you will later make another presentation at a job interview. Whether it is a departmental seminar, participation in a national meeting, an informal or formal speech, or a job interview, you must perform well when you are "on stage" to become recognized as competent and articulate, two qualities almost essential to your success as a scientist.

DEPARTMENTAL SEMINARS

A specific time, perhaps weekly or semimonthly, may be set up in your department for seminar presentations. Graduate students and other speakers will be scheduled to present information on their research, on the scientific literature, or on other subjects of interest to the department. During your presentation, keep in mind that you are talking to the people who know you best, but they are also the people who will or will not recommend you for career positions. Whether it is evident or not, professors evaluate your research and your communication skills any time you make a presentation in a class or for a departmental seminar session. Attendance at these departmental seminars is usually considered one of your professional or academic responsibilities, and participation offers you several advantages.

Seminars Provide Information on Current Research

For both the speaker and the audience, seminars present a unique educational opportunity. Every discipline includes a broad range of subject matter. You cannot expect to become proficient in all areas, but you can obtain some knowledge of and respect for the work being done in specialized areas other than your own. Conversely, when you present your own research to those in related but different areas, you provide them with another view of their discipline. Whether you are listening or speaking, seminars furnish a rather painless way to broaden your education.

Seminar Presentations Provide New Perspectives for Your Own Work

When you prepare and deliver a presentation, your perception of your own study increases. As Vernon Booth (1993) suggests, "Nothing clarifies ideas in one's mind so much as explaining them to other people." The learning that takes place in making your own presentation or observing others will improve your research. From others who discuss their research proposals, methods, and results, you will acquire ideas that apply to your own work. Seminars present a means for uncovering errors, picking up new perspectives, and strengthening your own research. Constructive, professional criticism is always beneficial for both the beginning scientist and the experienced professional.

Seminar Presentations Increase Your Ability to Evaluate Research

As a graduate student, you soon learn that neither the scientist nor the science is infallible. Much time, effort, and objective criticism are required to judge whether a scientific paper or presentation reflects good research, whether it is presented well, and whether it contains significant new ideas. The ability to listen to and critically evaluate a presentation is useful in acquiring new ideas and in deciding what information is valuable for your own research or communications.

Presentations Improve Your Ability to Communicate

Education and scientific progress are so closely allied with personal communication that everyone involved needs to develop an ability to communicate well. Few can become effective speakers without conscious effort. During graduate school, you will not make enough presentations to provide the desired training, but put forth your best efforts when you give a presentation and critically observe the efforts of your colleagues. The experience you gain will be well worth the effort at a professional meeting or a job interview as well as for career-long communication efforts.

THE PROFESSIONAL MEETING

Communication at scientific meetings transpires through both the spoken word and body language used formally and informally. The best information from meetings often comes from casual conversations. More formally, the speaker is a valuable part of the poster or slide presentation. When you are talking about science and research, you need to maintain your professional attitude whether you are in a formal or an informal situation.

The importance of chance encounters and casual conversations should not be underestimated, and some preparation can be made for these exchanges. Prepare for more than just an oral or poster presentation. Good impromptu communication requires that you know your own material and the literature on the subject. Know how your research was planned, designed, and carried out; how data were collected and analyzed; and how your results compare with those of others who have done similar research. Prepare for informal discussions by going over your material before you get to the meeting, and perhaps take notes with you. Try to predict what questions might be asked about your work. To get the most from a professional meeting, plan ahead for the activities you may be involved with.

Getting the most from a professional meeting:

—Study the program that is usually published well before the meeting, and plan your own schedule.
—Plan to give a slide or poster presentation. Enter a contest if one is held.
—Carefully select other presentations to attend, including those by authors whose publications you have read.
—Unobtrusively critique the good and bad points in posters and slide presentations to apply to your own research and communication.
—Observe the leaders of your society and how they conduct the meeting.
—Take advantage of placement or career services.
—Meet as many new people as you can. Join informal discussions about your research and that of others.
—Schedule time to relax and enjoy highlights of the town with your friends and new acquaintances.

Presentations at Professional Meetings

Relative to your career, the highlight of the professional meeting is your oral presentation or a poster. Both of these formats are prominent at meetings, and both require your communication skills. You will often be required to make the decision on whether to present a poster or a slide talk. Try to get experience with both formats. Consider characteristics and requirements for both and choose on the basis of which best fits your material and your ability. The comparisons and contrasts in Table 13-1 may be helpful.

As you are preparing your presentation, read Jay Lehr's editorial (reproduced in Appendix 13), and keep your audience's welfare in mind. You must coordinate the subject and your visual aids with the audience as you present your work whether to a full audience or a single poster observer. Remember that most of the people in your audience will know less than you do about your subject. Prepare your talk or poster for these people. Try to develop the clearest and most effective way to explain the subject to them. If you aim your presentation at the few people in the audience who know more than you about the subject, you may succeed in convincing them that you understand your material, but likely those few will not be impressed if you obscure the real significance of your subject from most of your audience.

TABLE 13-1
Comparison of Characteristics and Requirements
for Slide and Poster Presentations

Poster presentations	Slide presentations
The Situation	
Relatively informal; contact one to one or one to few	More formal; contact one to many
Both speaker and audience standing	Speaker standing and audience seated
No moderator; direct contact, no buffer between the speaker and audience	Moderator helps to introduce, buffer the audience, and keep time
Time limit flexible	Time limit formalized
Audience free; only truly interested remain	Audience more captured; most not likely to leave
Chiefly question/answer or conversational discussion	Chiefly declamation from speaker with a short question session
Handouts helpful; easy to exchange names and addresses	Handouts possible; less likely to exchange names and addresses
Preparation	
Materials: poster, tacks or loop/hook tape	Materials: slides or disc and notes
Know your subject—be able to justify objectives, refer to literature, and support your methods and results	Know your subject—be able to justify objectives, refer to literature, support methods and results
Prepare answers to likely questions	Prepare formal speech and slides
Get ready early. Construct poster, review and revise	Get ready early. Practice, review, and revise

Even with your own research peers, limit the use of statistical and technical jargon, but indicate what statistical analyses have been applied to your data. In most scientific research, statistical techniques are used only to provide a test of significance or to obtain an empirical mathematical expression of relationships. Emphasize the fundamental scientific concepts, not the statistical techniques. If you must use jargon terms peculiar to your subject, define them clearly for your audience.

Try to orient your talk or poster around **one central idea.** Accept that everyone in your audience will forget most of what you say. But if you do your job well, most of the audience will remember you and your point of emphasis for at least a few days, and those working in your area will remember far longer. If you fail to distinguish between big points and little ones, your audience will not make that distinction for you. An audience will simply walk away from a confusing poster and a confused author. The audience may sit in their seats as a matter of courtesy, but their minds will have turned to more interesting subjects. Also keep in mind that they are there because they are interested in what you have to say. Present your objectives and results vividly. Restrict the scope of your subject so that you can give a thorough explanation of the essential points.

For clear communication, you must be conscious of symbolic communication and communication without words (see Chapter 14). Your attitude, facial expressions, tone, and all the symbolic displays in your slides or poster may carry stronger messages than anything you say. Visual aids, properly prepared and used, can enhance presentations. At their best, however, visual aids are merely aids. At their worst, they can completely destroy the effectiveness of your presentation. They do not substitute for adequate preparation and effective verbal exposition by the speaker. For the poster they should support and illustrate the written material and your comments or responses to questions from viewers. For slide presentations, consider yourself and your speech content, not your slides, the central focus of the presentation.

SPEAKING AT THE JOB INTERVIEW

You are even more the central focus when you make a speech at a job interview. Taking a slide presentation to a job interview may appear more frightening than making other presentations, but it need not be. If you have taken every advantage of experience in departmental seminars and at professional meetings, you should have built the confidence that can make you a good speaker. Take that confidence with you to the interview.

For the speech at the job interview, two points deserve special attention and are unique for this situation: the **audience** and the **purpose** of the speech. For your departmental seminar, your audience is made up mostly of people you know, and many of them know what kind of research you have been doing. You are usually given enough time to fully explain your points. At the professional meeting, your audience is a group of people especially interested in your topic; they probably already know much about it. Your time is short, but they don't need a great many details to understand a point you want to make. Most of them are more interested in your research than in you. The same is not true with the job interview.

The **audience** at the job interview is often made up of administrators, managers, and scientists with diverse backgrounds who are interested primarily in finding out more about you, especially your communication skills. There may be few with any expertise in your specific area of research. You must present your material so that it is clearly understood by individuals who are probably highly intelligent but uninformed about your subject. Because the audience is different from those in your department and at a professional meeting, don't expect to make the same presentation to them that you used for the other occasions. You need to revise a presentation every time you present it to a new audience, especially when a job is at stake.

Because the audience's chief interest is you, you must align the **purpose** of your talk with this specific semantic environment. They probably have several qualified applicants. They want to know whether to hire **you**, whether you will work well with them, and whether your expertise fits the position they need to fill. They want to know if they would enjoy a professional association with you. Your purpose then is to provide positive answers to such questions. To do so, simply follow the principles for all good presentations, but alter your own approach to accommodate the different purpose.

Give your audience the opportunity to discover more about you than is evident in a research presentation. For example, after you are introduced, it is a good idea to leave the lights on for a minute or two, thank the audience for inviting you, and give a brief but not overly zealous explanation of your interest in the job. Then with a smile and a transitional remark such as, "My interest in this kind of work has increased with my research on…. Today I'd like to show you one part of that research in which…." At that point you are ready to adjust the lights and turn on the projector. This brief interlude between your being introduced and your presentation can put the concentration on you and can add immeasurably to the audience contact.

As with any presentation, maintain good eye contact throughout the talk. If you have a choice, keep the speech itself short, no more than 20 to 30 minutes; explain fully a limited number of points; and relate your study to that of other researchers. Some of them may be in your audience. Establish credibility with your experimental design and analyses, and report results for which you have strong evidence. Discuss possible meanings or applications for the results, and exude both confidence and humility. This is not the time for you to speculate on momentous breakthroughs that you believe you have made in science, although you should not be overly modest. Show the audience an example of the good work that you have done, and invite questions so that you can provide answers that will establish your expertise. Your attitude should be that of any good speaker—confidence flavored with a good dash of humility.

Limit the amount of material you present. The mistake I see made most often is that scientists seem to believe that they must display all the data they have collected and analyzed over a period of several years. Be selective; present only a limited but impressive part of your study. A few years ago, I watched a former student of mine interview for a position. He was given 30 minutes for his talk, and he unwisely decided to present work that he had done since he had received his doctoral degree, plus a segment on the different work he had done for the doctorate, and still another subject from the data collected during his master's degree. When I asked him why he had presented so many studies, he said that he believed the audience would be more impressed with the amount of work he had done than with the details. I disagree. The audience can assume that he has done a great deal of work in earning two degrees and holding a responsible position for 3 years thereafter. What they can't do is digest three complex studies in 30 minutes. The speech had loose organization to accommodate all three studies, it ran beyond 30 minutes, and there was not time to demonstrate credibility with details and to show quality research. (The young man was not offered the position.) Audiences are accustomed to time limitations. Establish your credibility, present quality methods and results, and show the relationship between your work and that of others. The audience can then readily assume that you have done other work of the same caliber.

Don't be surprised if, after your talk, the listeners seem rather uninterested in your research and ask questions that have no relationship to what you've been talking about. Remember that their concentration is on you, not your research. For them, your talk serves as a critical demonstration of what you can do and of how articulate you are. They already have a resume, transcripts, and letters that reveal your experience and abilities; they are now checking out a personality. Recognize that point and respond to it positively before, during, and after your talk.

One last word on presentations at job interviews. Do your homework. Learn all you can about the position, the location, and the people in your audience before you get to the interview. Information on the Internet or at the library can tell you about a company or university or agency to which you are applying. Often asking questions of the person who has invited you to come for the interview or the secretary who answers the phone can help you to understand the semantic environment in which you will be interviewing and making your presentation. Your major professor or other advisors can be helpful in providing background information and preparing you to go to the job interview. They may know some of the personnel you will be talking with. Knowing what kind of research is being carried out by particular scientists, even if it is unrelated to yours, can influence whether you are offered a job. But paramount to all this background knowledge are your own communication skills; present your best.

THE QUESTION AND ANSWER SESSION

In any formal presentation, your interaction with the audience is crucial to your success. This interaction is perhaps most focused during the question and answer session. The question/answer session allows the audience to clarify points or add to their knowledge of your subject and find out more about you. It can build your reputation as a scientist and speaker, and it provides you with an opportunity to surmise the strengths and weaknesses in both your research and your delivery by the kinds of questions asked and your ability to answer them. You must keep the entire audience in mind during the question/answer session. Preparation for the session requires that you know your subject and maintain your confidence.

Give clear, concise answers. Don't dismiss any question without a response, but don't belabor any point. Any question is important even if it sounds trivial. Don't allow yourself to be pulled into a controversy; although you probably know your subject better than most of the people listening, the time and place are not appropriate for any heated disagreements. After the presentation, you may want to continue a discussion with an individual, but do so only after you have released the audience.

Most people will not interrupt you during your talk, but if someone does, don't panic. Answer his or her question or respond to a remark courteously and completely but as briefly as you can. Keep your place in your own presentation (probably via your notes or slides) and return to your prepared speech as quickly and smoothly as possible.

Above all, maintain a professional attitude throughout the question/answer session. Many speakers tend to lose their professional demeanor when the last note on their conclusions dies down. They may loosen a tie or lean on a podium and relax their diction. "Yeah" is not a good way to begin the answer to a question. Avoid these distractions and maintain your role as speaker. If possible, let the moderator make the transition between your speech and the question/answer session. This technique gives the speaker a chance to relax momentarily. Anholt (1994) and Haakenson (1975) provide some good information on handling questions. The following suggestions may also help.

1. Listen closely. You cannot answer well without hearing and under-standing the question. Don't interrupt before the question is completed, even when you know what is being asked.

2. Repeat the question aloud if there is an even remote chance that it was not heard or is not clear to you or the audience.

3. Pause. There's nothing wrong with taking 2 or 3 seconds to think, and your answer will probably be better for it.

4. Answer the question completely but as briefly and directly as possible. Don't go into a new speech. Others may also have questions.

5. Take questions from various parts of the room. If the same person keeps asking questions or wants to discuss an issue beyond a reasonable

answer, suggest to him or her that you meet to discuss the matter further after the session.

6. Maintain eye contact most of the time with all the audience, not just the person asking a question. Although an individual is asking, you are answering to all.

7. Don't be afraid to say you don't know. Questions may be asked that are only remotely related to your subject. Simply indicate that your research has not supplied an answer to the question. Refer to the literature if you know a source for an answer, but don't guess. Never try to bluff an audience.

8. Reply courteously to all, and don't become defensive. Accept statements and "loaded" or trivial questions and maintain you professional composure. You can often dignify a question or comment that was not presented with dignity by supplying a serious, professional reply that is related to the subject.

9. Always maintain your dignity. Anger is the easiest way to lose it. The audience will have increased respect for you if you reply to the hostile question with a smile and a serious answer.

10. Don't speak beyond your time limit. End the questions if the moderator does not do so, and make a final summarizing statement if possible.

ROLE OF THE MODERATOR

Whether the occasion is a departmental seminar, a speech at a professional meeting, a job interview, or some other speaking situation, the speaker may need to coordinate efforts with other speakers, a program coordinator, a slide projectionist, and especially a host or moderator. The speaker should arrive at a meeting early and meet the moderator and perhaps the projectionist. If more than one speaker is on the program, the projectionist needs to know who you are and when your speech is scheduled so that he or she can have your slides or disc ready. Let the moderator have any information he or she needs to introduce you, and be sure to coordinate your efforts relative to lights, time signals, and the request for questions.

On the other hand, you may be the moderator and chair an entire session at a professional meeting. Be sure that you can pronounce the names of presenters and the words in their titles. Your job is to introduce them, help them feel comfortable, and solve or buffer problems that arise. In chairing a session, you should provide transitions from one presentation to the next, and be sure that you keep everyone on schedule so that one speaker does not encroach on the time of another.

To Be a Moderator

1. Obtain copies of abstracts or information about the talks you are moderating and familiarize yourself with each topic. Prepare a few relevant questions for each speaker to get the discussion started if the audience does not.

2. Help the speakers arrange visual aids or needed equipment. The moderator should coordinate the operation of lights, projectors, and other equipment with the speakers and should be present in plenty of time before the session to assist with last-minute details.

3. Talk with each presenter or obtain a brief resume and prepare a short introduction, including the speaker's name and title, academic and professional background, any special distinction, and title of presentation. At professional meetings, don't use up the speaker's time. Your introduction may include simply the speaker's name, institutional affiliation, and the title of the presentation.

4. Keep up with time. For example, with 15 minutes provided, we might expect the speaker to talk for 12 minutes (plus or minus 1 minute) and answer questions for 3 minutes. The moderator must be responsible for keeping everyone on schedule, that is, see that speakers start on time and finish on time.

5. Be sensitive to problems the speaker may have. Check equipment and know where replacement bulbs are located. Coordinate all efforts with the speaker. Buffer him or her from a hostile question or a string of questions that do not allow the speaker to move on.

6. Accomplish all your responsibilities in a congenial and professional manner.

FIT THE OCCASION

Scientific presentations can take numerous forms other than those described here for departmental seminars, professional meetings, and job interviews. In these situations, as well as presentations to such groups as public school children or civic clubs, you may make a speech without visual aids, provide a demonstration of a scientific reaction, host a video or film presentation, or serve as moderator for a symposium or a group discussion. Equipment may dictate the kind of visual aids that you use. Be sure to find out whether an electronic projector, a slide projector, or an overhead projector for transparencies is available, and be ready for any problems that might arise with such equipment. Always have a back-up plan; plan B has seen many people through difficult situations. Adapt to the situation, but keep basic principles of clear communication in mind as you make use of new situations and alternative media or equipment.

In the following chapters you will find more specific information on visual, verbal, and symbolic communication used in speech making, slide and poster presentations, and group communications. Most of the decisions on how best to communicate rest with you, but knowing the expected conventions can serve you well in making these decisions. Communication is both a personal and a social activity. Be creative, but also rely on standards or conventions that everyone uses. In other words, be yourself and use techniques that best serve your personality, but satisfy the expectations of your audience by using conventions that they will receive well.

In addition to Lehr's admonitions (Appendix 13), Anholt (1994), Booth (1993), Peters (1997), and Tierney (1996) have very good suggestions for making oral presentations. Briscoe (1996), Woolsey (1989), Anholt (1994), and Knisely (2002) cover the basic conventions for presenting posters. You can also find information on the Internet. For example, one good site on presentations is from Oregon State (2003) and can be accessed at http://oregonstate.edu/instruction/bb311/discussion4.html. Those sources, as well as this one, reflect the personalities of their authors and may not always be in agreement with each other or with you, but the principles of communication remain the same. Consider the audience, the subject, and the format and present information in the simplest, clearest way possible with your own personality and abilities.

References

Anholt, R. R. H. (1994). *Dazzle 'em with Style: The Art of Oral Scientific Presentation*. W. H. Freeman & Co., New York.

Booth, V. (1993). *Communicating in Science: Writing a Scientific Paper and Speaking at Scientific Meetings*, 2nd ed. Cambridge University Press, Cambridge.

Briscoe, M.H. (1996). *Preparing Scientific Illustrations: A Guide to Better Posters, Presentation, and Publications*, 2nd ed. Springer, New York.

Haakenson, R. (1975). "How to Handle the Q & A. " In *A Guide for Better Technical Presentations* (R. M. Woelfle, ed.), pp. 158–170. IEEE Press, New York.

Knisely, K. (2002). *A Student Handbook for Writing in Biology*. W. H. Freeman, Gordonsville, VA.

Oregon State University. (2003). "Data Organization and Presentation." http://oregonstate.edu/instruction/bb311/discussion4.html (verified June 6, 2003).

Peters, R. L. (1997). *Getting What You Came for: The Smart Student's Guide to Earning a Master's or Ph.D.*, revised ed. Farrar, Straus & Giroux, New York.

Tierney, E. P. (1996). *How to Make Effective Presentations*. Sage, Thousands Oaks, CA.

Woolsey, J. D. (1989). Combating poster fatigue: How to use grammar and analysis to effect better visual communications. *Trends Neurosci.* **12**, 325–332.

14

COMMUNICATION
WITHOUT WORDS

"It is the province of knowledge to speak,
and it is the privilege of wisdom to listen."

OLIVER WENDELL HOLMES

Hall and Hall (1990) believed that "90% or more of all communication is conveyed by means other than language, in a culture's nonverbal messages." The percentage may not be that high in communication in the sciences, but in any communication effort, we must recognize the transfer of information without, or in addition to, the words we use. All these elements influence the semantic environment. Listening and reading constitute much of communication, but we also see meanings with sight and touch. In speech, every gesture and every facial expression add dimension to what we say. With written and visual media, we also use language beyond the words. Technical editors are trained to use type styles and sizes, as well as other symbols, for emphasis to guide readers through a text. As authors design manuscripts for photocopy or for electronic publication, they become responsible for symbolic language that used to be in the domain of editing. Symbols, spacing, colors, and other embellishments can be as important as words in written and visual communication. In speech, physical circumstances, body language, and listening habits all contribute to success or failure of information exchange.

SYMBOLS

Conventional symbols are available, and new ones can be designed to help us express our organization of ideas and the precise meanings of terms. Conventional symbols for organization include the placement and size of headings, which constitute guideposts to lead a reader from one section of a paper to another. Still common is the paragraph indention. Although recently it has become rather fashionable not to indent paragraphs, this small space helps the reader move from one idea to another, especially if additional line spacing is not used between paragraphs. I contend that this organizational symbol should not be lost. Like a period or a comma, the indention is an unobtrusive guide for the reader.

We often accept many symbolic conventions without noticing them. We pause at a period, as we stop at a red light, without consciously analyzing why. A great deal of simplicity in communication would be made complex if such symbols were not universally accepted in language. Words themselves are actually symbols for sounds and meanings, and other symbols add dimensions to these meanings. We have become accustomed to italics to denote scientific names of species and abbreviations to designate units of measure. With careful use of such symbols, the audience will likely interpret them as you intend them.

Dreyfuss's (1984) *Symbol Sourcebook* is based on a databank of more than 20,000 symbols. Such symbols supplement all languages. Some, such as the skull and crossbones image that indicates poison, are recognized worldwide; others are limited to a small group of people or perhaps to specialists in a given academic discipline. Sign language, Morse code, chemical structures, braille, traffic signs, and hundreds of other symbols all carry important messages. Probably the most common symbols that many of us see everyday are the icons on our computers; communicating via the computer requires that we recognize their meanings. Symbols are handy tools for language if both the sender and receiver of a message interpret them in the same way.

Size, shape, spacing, color, and location as well as underlines and bold print can be used with discretion to help organize, emphasize, and clarify meanings in your writing or in visual displays for posters and slides. Without discretion, such symbols can simply confuse an audience. If every third word in a text is underlined, the underlining has no meaning beyond counting words, but if only one word in a paragraph is underlined, the reader interprets the symbol as suggesting emphasis. The use of large print, boldface, color, all capitals, white space, and positioning—all call attention to a spot or a word. For example, a headnote set in 14-point type, in total capitals, in boldface or italics, centered or underlined over an indented paragraph in 10-point type indicates that the headnote is the subject of that paragraph. For such a note to follow the paragraph or to be buried within it not only would lack the intended meaning but also would confuse us by interrupting the conventional arrangement of ideas.

All these points seem obvious, but prudent use of symbols can facilitate understanding, and confusion can originate from careless or imprecise use that might create false emphasis or ambiguity. For example, if you label items in a list as 1, 2, 3, 4, the reader may assume that the first item is more important than the second. Designating items with a, b, c, d may be less likely to draw that assumption, and using a consistent bullet or shape such as °, ➤, or * in front of each item is even less likely to suggest that the first item is most important. However, for complete clarity, you may need to tell your audience that the items are of equal importance or are listed in order of importance.

The overall appearance of the page or visual aid conveys a message even before the audience has read a word. Increasing the widths of margins to place a section of text inside the main text indicates that you are quoting another source of information or presenting an example or a supporting point. Successive indentions may indicate less importance or a dependence of the material with the deeper indention on that above it. The same is true of smaller type placed against larger type, and thick lines are more emphatic than thin lines. The possibilities are almost limitless for combinations of spacing, underlining, boldface, and other elements of emphasis.

Spacing and positioning of text and images can add to or distract from the intended meaning. Too many of these symbols used together are counterproductive or what Keyes (1993) refers to as "perceptual overload." Simplicity is the principle that should be applied to nonverbal signals as well as to verbal and structural elements in communication.

TYPE STYLES

Especially with word processing, slide making, and poster construction, you need to understand something about type sizes and styles. Printers have various styles of type that they refer to as "faces." We select a **font,** which consists of a size and face of type. Faces have names such as Times New Roman, Arial, Script, Old English, and many others. Size is measured in units called "points." What you are reading now is 10-point New Caledonia type; the titles for the chapters in this book are in 22-point New Caledonia. Another printer term that you may need to know is serif. A serif is an extension beyond the main body or shape of the letter. The small horizontal extension at the bottoms or tops of letters such as the *l* or *h* in Times New Roman are serifs. Type called sans serif is blocky without these tails or extensions.

This text is san-serif, 12-point Arial bold.

This text is serif, 12-point Times New Roman bold.

The two examples above take up different space horizontally, although both are 12-point type. Notice that the Arial seems to have thicker, wider letters, but is more compact. The vertical height in both lines are the same; it is this height that determines the point. Type faces and sizes can be important when you choose a font for a manuscript or a poster. Some are easier to read than others. Be conscious of type styles and sizes when preparing a paper for publication or reproduction by photocopy. A scientific poster or word slide can also be much clearer if the text is in a large, easy-to-read font. **Keep in mind that lowercase letters with only the grammatically necessary capitals are easier and faster to read than are total capitals.** See other examples of type fonts in Chapter 17 on posters (Table 17-1).

COLOR

A great deal of variety can be accomplished with type font and other nonverbal elements in communication without color. Color simply adds another dimension to communication. Dark type on a pale background (or black on white) generally evokes little response from the reader beyond the meanings of the words themselves. But insert a bright red word, and the symbolic meanings with color can elicit a complex and even an emotional response. As Imhof (1982) says, "the concept 'color' is ambiguous." Scientific analyses of color do not take into account the emotional response a viewer may have to color. Because of the ambiguity in reactions, one can hardly provide a set standard for communicating with color, yet it is an important element, especially in scientific slide presentations and posters. As with other tools in communication, we need to understand basic conventions or expectations that are acceptable to an audience.

Imhof's (1982) theory of color provides a valuable guide for cartographers that we can apply to other scientific communications (see Appendix 12). He maintains that "A color in itself is neither beautiful nor ugly. It exists only in connection with the object or sense to which it belongs and only in interplay with its environment." So that red means one thing on an apple, another on a football jersey, and still another on an exam a teacher has just returned to you. In addition to the effect of environment on meaning for color, one's own personality, past experience, or the culture in which he or she lives may give to a color a particular meaning in a given context. Dreyfuss (1984) notes that each color has both positive and negative associations in various situations and cultures. People call colors harmonious or clashing. We speak of cold colors and hot colors. Blue is cool or cold; yellow is warm or hot. There are hot pink and cool green, earth colors in shades of brown and orange, and neutral grey, which some call drab and others beautiful.

Tenner (1996) suggests that "the eye can distinguish more shades of grey than of any other color." And in a feature in *The Wall Street Journal*

(17 November 1993), Laura Hays noted that the medical community finds "grey and scales of white to black as the safest and most practical color choice for digital images." Imhof (1982) says that "grey is regarded in painting to be one of the prettiest, most important and most versatile of colors. Strong muted colors, mixed with grey, provide the best background for the colored theme." This remark from a color expert may provide an important message for us when we choose background colors for posters or slides. See Appendix 12 for more of Imhof's discussion about color.

These opinions about grey do not mean that we simply use black and white and shades of grey for all our communication. Alone grey can be beautiful, but probably its leading role is to mix with other colors to subdue the tone. Brown can also subdue the brightness of some colors well. We can look around us and see the color preferences for cool, subdued backgrounds that humans generally seem to prefer to live with day to day. Drive down almost any street, and you will see that houses are typically not bright, hot colors but white or off-white, grey, tan, or brown tones. Even what are called red or yellow brick homes are more brown than red or yellow. And typically, the walls inside those homes are not bright, but pale subdued colors. Look across a parking lot and note how many of the cars are either black or white or any number of subdued colors—greys, blues, greens, browns, reds—but very few you see are bright yellow, bright red, bright blue, or bright green. Surely we can conclude along with Imhof that people prefer background colors that are subdued with grey or brown. If your car is the bright yellow one, that is fine; it reflects your personality. But in communication of science, consider the audience's typical preference rather than your own.

Care with the use of color is important in scientific communication. As Tenner (1996) notes, the prominence of some colors can be deceptive, and in her *The Wall Street Journal* (17 November 1993) feature, Hays says, "certain colors can look bigger or smaller in the same-sized area." Dreyfuss (1984) says "color produces immediate reaction and is the exclamation point of graphic symbols, so it must be reckoned with." He also states "one's attention is often captured by color before form or composition is completely distinct." With color's ability to attract, it **does matter** what colors the scientist selects for communicating an idea. Colors do have meanings that can either support or deter communication. For color selection, we rely on the natural environment for some meanings and on established customs in a culture for others. Hot pink would seldom be selected over a cool blue to indicate a geographical body of water or even as a background color for scientific slides.

The meanings of colors is often associated with safety. Electricians communicate with colored wires, and highway departments communicate important issues with color. Stop signs are red, caution signs yellow, and information signs blue or green along the highways. We have established that for traffic lights red means stop, green means go, and yellow indicates caution. When you use color to convey meaning, be sure to remember your color-blind colleagues. To them the top traffic light means stop and the bottom one means go. Position, spacing, size, or design may carry your message as well as color in such things as graphs and will not leave out these audience members.

For slide making, today's film recorders and computer programs are a godsend, but keep them under your control. Such slide makers can produce millions of variations in colors; such technology is far beyond the simplicity needed in scientific communication. Humans can hardly distinguish that many shades of color. Don't ever let color or any other symbolic ornamentation distract from the scientific message. Too many colors, large areas of strong or bright colors, inconsistent use of colors to represent the same meaning, or poor choice of color can be detrimental to communication. Selections for combinations of colors can be very important for slides and posters. For example, red text may contrast well and be quite readable on a white background, but it may not be clear against blue. If dark, subdued backgrounds are chosen for slides, the text should probably be a soft, clear, light color such as a not-too-bright yellow or white.

As with the meanings of words, the scientist must select a color that is aesthetically pleasing but will not distract from the science by calling attention to itself. Ask for the opinion of reviewers before you make final decisions on colors to use. Imhof (1982) says, "Subdued colors are more pleasing than pure colors." He goes on to provide six rules applicable for map design that may be equally applicable for other scientific presentations. I strongly recommend that you read Imhof's chapter "The Theory of Color" (partially reproduced here in Appendix 12), and when you are making graphs or other illustrations, consult Briscoe (1996), Tufte (1983), and Council of Biology Editors, now known as Council of Science Editors (CBE, 1988), for details on communicating with design as well as color.

PHYSICAL COMMUNICATION

In spoken communication, the physical setting in which you talk, your physical presence, and your body language are all a part of the semantic environment and can be more important than any words you say. If you have control over the **setting,** make it as comfortable as possible for you and your audience. If you have no control, be conscious of such things as lighting, noise, temperature, the arrangement of chairs, or other attractions and distractions. Sometimes you will have to make decisions between the lesser of two evils; for example, choosing between a comfortable temperature and a noisy furnace or fan. Your words are worthless if your audience cannot hear them or are too distracted by other elements in the environment.

The value of your words is also enhanced or diminished by your own physical presence and your **body language.** As Smith (1984) says, you "are your most important visual." You can also be your own greatest distraction. Every movement you make, small or large, contributes to your communication. You talk with your eyes, with your feet, with your posture as well as with your hands. Your grooming is one of the first messages you present to your audience. It is idealistic to suppose that the audience will "pay attention to what I have to say, not how I look." What you should wear depends on what your audience expects. The way you style your hair is certainly your own business, but it can still convey a message. Be attentive to your grooming. Styles do change, but conventions in clothing and appearance are relatively stable in the scientific community. Your grooming and body language reflect a tone or attitude that cannot be hidden behind words.

Physical expressions are as important as grooming; many expressions are extensions of your personality. You should not change your personality for an audience, but you can condition much of your body language within the confines of your personality. Often we are not conscious of our own mannerisms. One student had a habit of unwittingly batting his eyes as he spoke, and in making speeches for my scientific presentations class, the constant blinking was most distracting. We were able to call his attention to the habit, and with conscious effort he controlled it and was a much better speaker with that single change. Probably nothing is more important than eye contact, whether you are talking with a single individual or a large audience. For some cultures the focal positioning of your eyes carries very important meaning. In the United States, contact between the eyes of the speaker and those of the audience is expected to help convey a message. But you can also say things with your feet or your hands, a shrug of the shoulders, or a wave of the arm.

You may inadvertently emit distracting vocal sounds that clutter your message. Grunts; "ums," "ahs," and "you knows"; a constant clearing of your throat; sniffing; or other audible distractions can cloud your message. Some of us don't have the most beautiful voices in the world, and a physical condition such as a swollen adenoids or a cold may make it essential that we clear our throats, sniffle, or blow our noses, but control these things as much as possible.

Such control may not always be possible. Don't despair if you have a speech impediment or a physical condition that attracts attention to itself but is out of your control. One of the most successful people I know stutters—not just an occasional repeat of an explosive *B*, but repeated interruptions with stuttering over words or letters. He has not allowed this condition to interfere with a career that demands public speaking and group communication. When his voice hangs onto a letter, he simply allows the stutter to run its course, takes control again with no apologies to anyone, and moves on with his speech. Audiences respect him. The first-time listener may be momentarily taken aback, but in almost no time that person is listening to what is being said and, along with the speaker, essentially ignores the stutter. If this speaker exhibited added nervousness or an embarrassing self-consciousness, the audience would also hang onto the impediment and lose much of the speech content. By example, the speaker simply guides the audience in how to react. If you have an uncontrollable voice problem, a physical deformity, or even a temporary bandage on your face or hands, you need not try to hide the appearance or call attention to it by apologizing or appearing embarrassed. Almost everyone has some sort of problem. An audience will notice if you are extremely beautiful, tall, short, thin, or fat; have to walk with hand crutches; or sit in a wheelchair. Accept that fact, and make your listeners as comfortable as possible by not calling attention to whatever it is that makes you different. Whenever you can, control your appearance as well as the physical situation. For a scientific presentation, never make a point of attracting attention to yourself rather than your message.

Features of your voice and body language can have a very positive effect on what you have to say. Voice inflections, volume, and tone can change the entire meaning of words or make a point more or less emphatic. Practice to make your voice quality help to carry your message and coordinate the voice with the body language. A hand gesture, closed fist, open palm, or directional movement of the hand can help to clarify meaning. Facial expressions can indicate beyond the words whether you are serious or jesting and whether you believe in what you are saying. Ironically, clever use of body language along with meaningless jargon can sometimes make an audience believe you have said something important. Perhaps politicians and sales persons often depend on this ability. As a scientist, you should not mislead audiences to believe what you say is important when it is not, but spoken communication is made most effective with use of physical expressions to reinforce the words.

Position, posture, and space are also important in communication. Where you stand, how confident and erect you stand, and how close you stand to an individual or to an audience convey messages. Humans are territorial; to get too close may invade their space, especially between individuals. However, a step forward toward a group suggests bringing the message closer or including them more fully. We often move a step or two toward an audience when we invite questions. To enhance the semantic environment further for audience questions, be sure the lights are turned on the audience as well as the speaker. These techniques suggest to the audience that they are involved and their questions are welcome. On the other hand, to leave the audience in the dark or in subdued lighting or to back up too far or to plant yourself too firmly, similar to a speaking statue behind a speaker's stand, is less inviting to the audience. The convention of the speaker's stand or, perhaps better, simply of the speaker standing in front of a seated audience is important in relegating the speaker and listeners into their respective roles. The physical setting combined with the appropriate body language can produce a relaxed atmosphere for communication with an audience.

Your personality, cultural background, physical condition, needs, or beliefs can influence your mannerisms so that you are not conscious of your facial expressions or body movement. Ask an observer to point out the strengths and weaknesses in your body language. Control the expressions that enhance your scientific message as well as those that distract. Don't let your meanings get lost among physical attractions or distractions.

LISTENING

Body language and voice, more than the words themselves, motivate an audience to listen or not to listen. We often underrate the importance of listening. Half of the responsibility for spoken communication should be that of the listener, yet we design full courses for teaching speech and give little or no attention to teaching people to listen. Listening is a matter of hearing, observing, and thinking all at the same time in order to have the clearest perception possible of what a speaker is conveying. In your career you will likely spend more time listening than speaking. Use this time wisely. Many of us have bad listening habits, but they can be improved with practice.

Perception or comprehension, not just hearing, is the end objective. In fact, the ears are only a part of listening. If the speaker is physically present, the listener's eyes must watch the speaker and be sensitive to his or her physical expressions as well as the words voiced. Be sure your brain is engaged with the ears and eyes in order to avoid misinterpreting what is said. Be selfish; remember that you receive more from listening than from talking. To listen well, one must put aside any prejudices about the speaker or the subject and interpret objectively without any preconceived ideas. The good listener will be sensitive to all the elements in the communication. Pay attention.

It takes concentration and energy to listen well. As with other forms of communication, you have to maintain the right attitude. This attitude should be receptive and open-minded, not defensive or patronizing. Many of us have expended a great deal of energy and concentration in our lives on appearing to listen when we are actually tuning things out. We had other things on our minds when parents or teachers were talking, but we developed an ability to hold the proper facial expression and look at the speaker while our minds wandered. Or rather than paying attention to what is really being said, our thoughts are planning a reply to what we think is being said. We may hear the beginning of an idea and prematurely form an inaccurate opinion of the speaker's meaning. Or we may prematurely judge the speech content by a speaker's appearance or reputation; highly reputable scientists can make bad speeches, and novices can make outstanding ones. We may also engage in other mental or physical activities that distract from the listening. Just half listening or "listening with one ear" is as bad as mumbling or just half speaking. Listeners must do their part to make communication a success.

Listening in a small group in which ideas are being exchanged is probably the most difficult listening challenge. You must be able to change your focus quickly from one personality to another and synthesize or recognize relationships in the material from several points of view. Listening as a single receiver of information or as a member of an audience requires variations of the same good habits. A conscious effort on your part could lead to better listening. Try the following:

1. **Look at the speaker. If you need to take notes, don't let that action make you miss the nonverbal communication.**

2. **Be sensitive to the tone and inflection in the voice. Watch how body language and voice blend into meaning.**

3. **Don't interrupt. A speaker may be struggling for words that you could easily supply, but let him or her finish an idea before you inject a response.**

4. **Be receptive and concentrate on meanings. Relate what the speaker is saying to what you already know and thereby increase your knowledge.**

5. **Tune out the distractions. No setting is perfect. There will be noise, poor lighting, something else on your mind, or another member of the audience making distracting sounds or movements. Focus on the speaker and not the distractions.**

6. **Watch your own attitude and try to empathize with the speaker. Be open-minded. Don't jump to conclusions or judge the speaker by his or her appearance or accent, and evaluate what is being said only after you have heard the entire speech.**

7. **Interrupt with a question only if that question can't be postponed. When the speaker is finished, be sure you have understood meanings. Then, ask questions or paraphrase what was said to clarify any nebulous points.**

8. **Be sure to respond. In a small group or in one-on-one conversation, you can use your facial expressions or other body language to indicate when or how you are interpreting what is being said. You can ask questions to lead the speaker toward clear explanation. And when the speaker is finished, you can make your comments or observations. In a large audience, you may think your response is not important, but it is. Even if you cannot ask questions or make comments, your attention, posture, eye contact, and facial expressions are important to the speaker.**

9. **Be patient.** You can hear much faster than the speaker can talk, but remember listening requires more than just hearing. Use that extra time to think about what the speaker is saying and to make your nonverbal response.

10. **Don't be distracted.** It is demoralizing to a speaker to say, "Go ahead, I'm listening" when you are obviously being more attentive to someone or something else.

Listening, as well as the more tangible elements in communication without words, is crucial to clear understanding. Not just in visual displays but with any written or spoken communication, the value Imhof (1982) gives to "the composition as a whole" is fitting. Your manuscript or your speech is not just organized words and ideas. It also contains symbolic images, whether they are created by type size and style, color, physical circumstances, or body language. Concentration by the listener and the reader is also tantamount to successful communications. Pay close attention to your words and how they are organized, but also be conscious of the appearances and the listening habits of both you and your audience.

References

Briscoe, M. H. (1996). *Preparing Scientific Illustrations: A Guide to Better Posters, Presentations, and Publications*, 2nd ed. Springer, New York.

Council of Biology Editors (CBE) (1988). *Illustrating Science: Standards for Publication*. CBE, Bethesda, MD.

Dreyfuss, H. (1984). *Symbol Sourcebook: An Authoritative Guide to International Graphic Symbols*. Van Nostrand Reinhold, New York.

Hall, E. T., and Hall, M. R. (1990). *Understanding Cultural Differences*. Intercultural Press, Yarmouth, ME.

Imhof, E. (1982). *Cartographic Relief Presentations*. Walter de Gruyter, New York.

Keyes, E. (1993). Typography, color, and information structure. *Tech. Comm.* **40**, 638–654.

Smith, T. C. (1984). *Making Successful Presentations: A Self-Teaching Guide*. John Wiley & Sons, New York.

Tenner, E. (1996). *Why Things Bite Back: Technology and the Revenge of Unintended Consequences*. Alfred A. Knopf, New York.

Tufte, E. R. (1983). *The Visual Display of Quantitative Information*. Graphics Press, Cheshire, CT.

15

VISUAL AIDS TO COMMUNICATION

"Blessed is the man, who, having nothing to say,
abstains from giving us wordy evidence of the fact."

GEORGE ELIOT

As with other symbolic images, visual aids can complement or supplement words in both written and spoken texts. Their basic function is to clarify points beyond what the spoken or written word alone can do. In scientific communication if they attract attention only to themselves and do not increase clarity, they serve as distractions. Used effectively, they will strengthen the scientific message. Attention to the following five principles will help you avoid pitfalls in the use of all kinds of visual aids:

1. Make them simple enough to be comprehended easily.
2. Make the images or letters large enough to be seen clearly.
3. Make a trial run long enough before the real presentation to permit you to change the visual aids if they do not serve you properly.
4. Coordinate them carefully with the speech or text so that the audience is not distracted and the visual aid is relevant to the point in the speech or written text.
5. Just before a presentation that incorporates visual aids, be sure that any needed equipment is in place and operating smoothly.

With scientific presentations today, we generally think of visual aids as slides that accompany an oral presentation. However, all sorts of visual aids can serve the scientist in demonstrations, teaching, and presenting displays such as posters. A visual aid may be any visual image, including speakers themselves, their personal appearance and body language (Smith, 1984), or such things as a live cockroach to show its anatomy or behavior, a complex physical model of a process, or a drawing of a molecular structure. Although slides and electronic images projected on a screen and poster displays are probably the most common visual aids for scientific communication, other forms are still valuable in many situations. Images on paper (graphs, flow charts, structures, and others), chalkboards, flipcharts, videotapes, overhead transparencies, or the props themselves can serve to assist a speaker or writer in making a point.

Each of these kinds of exhibits requires its own equipment whether it be chalk and a chalkboard, a projector, or a poster board. Recognize the values and limitations of each. Flipcharts and posters accommodate small audiences and require well-lighted conditions. Slides require subdued lighting but can serve a rather large audience. Speakers often fail to note how handy a chalkboard can be. You should not write lengthy messages on it because your back is turned to the audience for too long, but without other visual aids or in answering a question, it is often helpful to quickly sketch a chemical structure or write an unusual word for the audience to view. Videotapes or filmstrips interrupt the speaker entirely but can be coordinated with a speech if they are relevant. I attended a presentation once in which the speaker introduced his subject, which had to do with the processes involved in germination and seedling emergence. After the introduction, with equipment ready to go at the push of a button, he showed a short time-lapse filmstrip of a soybean seed developing into an emerged seedling. He then provided a transitional remark and began his presentation of slides. The filmstrip was quite effective and did not distract from the speech or the subsequent slides. But two media can be difficult to coordinate, and interruptions to the flow of the speech can be distracting; be sure they fit smoothly into your presentation. Objects for show-and-tell can also be valuable or distracting; any prop needs to be big enough for all the audience to see and yet not ostentatious enough to distract throughout a presentation. The size of the audience can be an issue. Generally, passing an object around the room is also distracting; it may be better to invite the audience to stop by to view the display item after the speech. The important point with any visual device is to make it serve, not obstruct, the communication.

Visual aids can add information, they can illustrate or provide examples or evidence, or they may repeat what you are saying or writing. In the poster, you can present information via a graph or photograph, and the written text will just call attention to that information. Similarly, in a slide presentation you may describe in words the effect of a chemical compound on a petunia plant.

Your words will be strengthened if you have a picture of a treated petunia to illustrate your point. In writing you have been taught not to be wordy or redundant. In speaking, repetition used discreetly is often the best way to strengthen a point. A reader can return to a point in a text and read it again. Because a listening audience does not have the option to listen again, a judicious speaker will often reinforce with repetition. Visual aids can help you duplicate a point in a second medium without appearing redundant.

Visual aids can also help compensate for language barriers. When your first language or your accent is not the same as that of your audience, the importance of supplemental aid increases. It is especially important that you display key words you have trouble pronouncing or those that are new to the audience. With this assistance from your visual aids, if you will speak slowly and enunciate carefully, an attentive audience will understand you. I am not referring just to international students or to professionals from non-English speaking countries who join your societies. The English-speaking world includes numerous accents. Whether you are from Burundi, Bangkok, Boston, Baton Rouge, or Bath, your accent will be foreign to someone in your audience. Along with being proud of the way you sound, be sure you provide any assistance that your audience may need for understanding you. The best advice I can give is **slow down** and provide informative visual aids.

As with other symbolic communication, the style, layout, color, type, and form for a visual image should be clear and aesthetically pleasing. Characteristics of symbolic media may be as important as is content in carrying messages to the audience. For example, color in a slide set or a poster can be a point of unity. Choose basic colors with care, limit the number of colors used, and coordinate them carefully with each other and with the content. Subjects, ideas, or transitions can be coded by color. For example, if you use a series of bar charts and, in the first, copper sulfate is represented with a blue bar, then copper sulfate should be represented with blue in any other image used. In addition to color, other characteristics essential to the success of a visual aid include type, consistency, size, form, density, spacing, and style. Be creative but also use conventions that the audience expects. Computers can provide many advantages in creating visual aids, but they can also produce complexities or outright errors. Use their capabilities carefully. In this and other chapters, I discuss primarily visual displays used in slide or poster presentations, but the same qualities are important for any visual aid you use.

SLIDE COMPOSITION

We have come a long way in our ability to make slides over the past 30 years from black-and-white reverse image photos to the use of vericolor film, diazochome development of film or colored gels. Many of you now probably don't even recognize what I'm referring to and may even consider 2″ × 2″ slides that sit in a carrousel as old-fashioned. The use of electronic projection is becoming the norm. I suppose the word *slide* should not be applied to electronic projection, but I find that the word has been transferred to the electronic images, and I will use it to mean the image projected onto a screen. As with other developments in tools for communication from the pencil to the typewriter to the computer, this electronic medium is a valuable advancement. But with any tool, dangers accompany the advantages, and we must be mindful of the communication principles of clarity and simplicity in our use of any tool. For example, the ability to animate slides electronically offers a great advantage of presenting material in a sequence that simplifies and keeps audience attention on a single point or lets us build a flow chart or other image piece by piece to facilitate audience understanding. But the overuse of animation or use that simply attracts attention to itself and not to the scientific point can be most distracting. When in doubt, don't animate or add other embellishments to slides that can divert attention from the scientific point. Use such tools to clarify an issue, not to distract.

My discussion here primarily concerns the principles for good communication, not the tools used. In referring to slides, I mean either those you put in a carrousel or those images projected electronically through a computer system. The software I am most familiar with is Microsoft PowerPoint®, and so I will allude to that software, but I am not recommending it over any other that you may wish to use. Others may be as good or better; the principles remain the same. Any single visual aid should meet the following criteria:

---

1. Simplicity—Usually one point with limited subpoints
2. Visibility—Distinct and legible to any person in the audience
3. Unity—Cohesive and uniform with other visual aids and with the written or spoken words
4. Quality—Clear, attractive, and aesthetically pleasing
5. Feasibility—Production and presentation possible with the materials, facilities, and time available

---

With few exceptions, the scientific slide presentation will consist of word slides, photographs and other illustrations, and data display in figures and tables. A mix of these kinds of slides is more interesting than all of one (see Appendix 13). Photographs can provide information, transitions, and visual relief in a scientific slide set that has a great many word slides or tables and figures. As you pursue your research, keep a camera handy; images from a digital camera will be easier to put into an electronic slide set, but a scanner can also serve to transfer prints. You will one day need pictures of equipment, plants, animals, symptoms, or other images from your research for written, oral, and poster presentations. Some may be as important or more important than words. Microphotographs may be the most informative evidence in the results of a study on a science that is invisible to the naked eye. For a slide or poster presentation, color photos are usually best, but keep in mind that you may wish to publish a picture with a journal article and that journals often publish only in black and white.

Courses in photography and numerous books for beginners are available. Plotnik (1982) has a good chapter on basic photography with a 35-mm camera, and you can apply the basic principles to a digital or any other camera. But you can learn a great deal on your own. If you have not had extensive experience with photography, acquire a camera and give yourself a short course by reading instructions packaged with the camera, ask questions of people at camera shops, and take pictures and study the results. As you take photographs, visualize the final product; the image is in your viewfinder. Be sure that the subject you want to display dominates the picture; some people take pictures too far away from their subject and include too much or distracting background. Certainly background is important. The photograph will appear two-dimensional, and tree limbs may look as though they are growing from people's heads. Placing your subject against a solid background, the sky, the earth, or an improvised backdrop can help. Lighting is influential in both outdoor and indoor photography. Outside, the sun may be too bright, or it can cast heavy shadows that will subdue your subject. Watch the depth of field, and be sure your subject is the area in focus. Compile a file of photographs associated with your research project; they will be invaluable when you begin making slide sets and posters.

Perhaps most of your slides will not be photographs but text, for example, title slides, key words, and data in tables and figures. Whatever method you use for producing word slides, the resulting image will have relative dimensions of two units by three units. Try to keep the orientation horizontal, that is, the longer dimension will run across the screen. Balance the content on the screen with select words or symbols that are spaced appropriately and clearly visible from anywhere in a large room when projected onto the screen.

Plan your slides for viewing in a large room, and you can use them in a small area as well. Generally for Microsoft PowerPoint® images, the smallest text on the slide should be no less than 28 points. Most of the text should be about 32- to 40-point type with limited serif or sans serif, and titles and headings can be 60-point or greater. I suggest such typefaces as Helvetica, Arial, Tahoma, or similar styles of sans serif or Times New Roman if you like serif. A general rule of thumb is that a slide that is clearly legible without projection at arm's length (approximately 25 inches or 0.65 m) can be read with projection onto a screen at a relatively long distance (approximately 50 feet or 15 m). Each slide should make one basic point with perhaps a subpoint or two, and the text on the slide should cover no more than 14 lines (10 or less is better). Lines should never be longer than 40 spaces (34 or less is better). Note the following:

> **Objectives**
>
> 1. To compose attractive, legible visual aids
> 2. To conserve both time and cost in slide production

This example is composed of eight vertical lines, including the blank ones, and the longest line contains 25 characters. It will fit the two-by-three format for slides, and the 16 words are not too many for an audience to comprehend quickly, especially if you direct their attention to each objective with a pointer or animate them into view one at a time. More words or lines would make the slide heavy.

Fill the screen but not with too many words or ideas. Text for word slides should be limited to key words and phrases, not lengthy sentences and paragraphs. The speaker should fill in the information needed about those key elements. Complex slides can be made more immediately comprehensible by breaking them down into individual points or parts so that an audience can follow the speaker through the complex information. See the examples in Appendix 13 for designing a single word slide or for building a slide with a series of simpler parts (the house-that-Jack-built technique) as well as a full, effective slide set. Anholt (1994) has good descriptions and examples of simplifying complex slides.

In addition to spacing and sizes, in slide making be careful in combining colors or using a colored text on a colored background. For instance, a bright yellow may bleed into a blue background or cause a blurry whitish image at the interface. Red may appear sharp and distinct when used on a white background, but on blue it may appear dull and subdued and be difficult to read. Use solid distinctive colors for such things as bars in bar charts but choose them carefully so that one bright color does not appear more important than others. Make trial runs with various colors of backgrounds and text or bars and lines to provide model slides to base your color selection on. Then use consistent colors for backgrounds and texts throughout a given slide set. Keep in mind that the light intensity on the projector will differ from that on your computer monitor, and the colors will not be the same when projected. Choosing dark subdued backgrounds and contrasting light colors for the text is a relatively safe choice if you can not try your slide set out on the projector before your presentation. Further comments on the use of color are in Chapter 14 and in Appendix 12.

These principles on size, spacing, and color are especially true for tables and figures. For a slide presentation on research, your data are the most important support to your objectives and conclusions. Present only the data points that are essential to your talk. More than 20 items in the field of a table are probably too many; fewer than 16 would be better. In a line graph, three lines are much better than five, and more than five can be too many. In a bar chart, grouping can affect the number used. In groups of fours, 12 bars may be understandable; in groups of twos, 8 or 10 bars are acceptable; but for single, separate bars, try to limit the number even more. Let exceptions to these standards be rare. Every added bar or line adds to the complexity for the audience. You may be able to divide tables that might be published as single tables into two or more by separating variables. If comparisons of these variables are needed, then use the house-that-Jack-built technique described in Appendix 13.

Often you can reduce the number of data points in tables and figures by recognizing that representative data will illustrate your points and you can verbally discuss any omissions. If three different lines in a line graph run along similar data points, take two lines out and tell your audience that the data omitted are similar to those represented by the line on the slide. This technique is illustrated in Appendix 10. Recognize that requirements for data presentation with a speaker there to explain are not the same as for publication. Readers can study a complex table or figure for as long as they like; viewers must comprehend the point immediately in order to pay attention to what the speaker is saying before the visual aid is gone. **Never expect simply to transfer tables and figures from a publication to visual aids for a slide presentation.**

Notice that Table 15-1a could appear in a publication, but it should not be made into a slide for two reasons: (1) the type would be too small for an audience to read, and (2) if they could read it, it contains too many numbers for them to comprehend in the short time a slide should remain on the screen. To produce an acceptable slide, choose representative data that best illustrate the point you wish to make. For Table 15-1a, let us assume that you wish to emphasize differences in the total nitrogen. Note how Table 15-1b accomplishes this purpose and preserves an acceptable type size with the limited number of data points. As a speaker, you can point out any differences that you need to discuss relative to shoots and roots, or if you have time for the specific discussion on those points, you might produce two additional slides with that information. Three slides with limited information communicate better than one that is too heavy. The speaker also can provide the additional information that was in the original caption and footnote without putting it on the slide itself. Table 15-1b has no table number and no footnote; the number is superfluous in a spoken presentation, and the speaker can explain that the letters represent statistical significance. Notice that the same limited data from Table 15-1b can be displayed in graphic form, as in Fig. 15-1. Unless you

TABLE 15-1a
Table from a Thesis
Table 4. Influence of *Bradyrhizobium japonicum* USDA 110 (BR) and *Heterodera glycines* Race 3 (SCN) on Root, Shoot, and Total Nitrogen Contents of 'Lee 74' and 'Centennial' Soybean Cultivars.

Soybean cultivar	Treatment	Root nitrogen %	mg/plant	Shoot nitrogen %	mg/plant	Total nitrogen %	mg/plant
Lee 74	Control	1.84 a*	11.5 c	2.46 bc	34.5 c	2.26 ab	46.0 c
Lee 74	BR	1.85 a	16.5 abc	2.94 ab	85.9 ab	2.68 ab	102.4 ab
Lee 74	SCN	1.78 a	14.3 bc	3.49 a	78.5 abc	3.04 a	92.8 abc
Lee 74	BR + SCN	2.03 a	21.0 a	3.17 ab	96.7 a	2.86 a	117.7 a
Centennial	Control	1.57 a	14.4 bc	1.95 c	38.7 bc	1.82 b	53.2 bc
Centennial	BR	2.12 a	17.9 abc	2.55 abc	64.6 abc	2.44 ab	82.5 abc
Centennial	SCN	1.67 a	17.7 abc	2.52 abc	65.2 abc	2.28 ab	82.8 abc
Centennial	BR + SCN	1.83 a	23.6 a	26.9 abc	87.8 a	2.44 ab	111.4 a
	LSD ($P = 0.05$)	0.75	7.1	0.97	48.3	0.88	53.8
	CV(%)	23.59	24.0	20.63	40.4	20.59	36.1

*Means within a column followed by the same letter are not significantly different at the 5% probability level. Data are means of 3 replications.
From the thesis of David Mersky (University of Arkansas, Fayetteville, 1992: p. 40). Used with permission of the author.

TABLE 15-1b
Adapted Table

Influence of *B. japonicum* USDA 110 (BR) and *H. glycines* Race 3 (SCN) on Nitrogen in 'Lee 74' and 'Centennial' Soybean.

	Total Nitrogen (mg/plant)	
Treatment	Lee 74	Centennial
Control	46.0 c	53.2 bc
BR	102.4 ab	82.5 abc
SCN	92.8 abc	82.8 abc
BR + SCN	117.7 a	111.4 a

From the thesis of David Mersky (University of Arkansas, Fayetteville, 1992: p. 40). Used with permission of the author.

need to present specific amounts, this visual image may be more effective than the table.

Some equipment allows for more flexibility than do others, but you, not your equipment, are responsible for your slides. Make the equipment work for you. Note the following suggestions for composing word slides as well as tables and figures:

1. Use a thick, blocky type. Italics, script, or thin letters are harder to read.

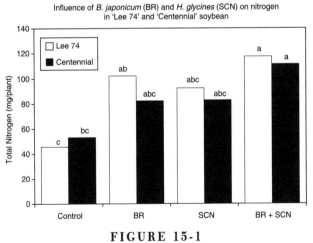

Influence of *B. japonicum* (BR) and *H. glycines* (SCN) on nitrogen in 'Lee 74' and 'Centennial' soybean

FIGURE 15-1
Graphic display appropriate for a slide.

2. Use lowercase letters, capitalizing only where grammatically necessary. Lowercase or a mix of lowercase and uppercase is easier to read than all capitals.

3. Remember that projection will expand the spaces on the slide as well as enlarge the letters. Usually use single space between lines of continuous text and 1.5 or 2.0 space only between ideas or to set apart a caption or heading.

4. Remember that the audience should understand the point on the slide immediately, and don't overload it with information or visual distractions.

5. Ordinarily, don't number your tables or figures. You may want to rearrange your talk and put Table 3 before Table 1. With slides in sequences with the talk, the numbers are superfluous.

6. In typing tables, space evenly between columns. Always align decimals.

7. Be sure lines and symbols in figures adhere to the same standards as text. Thin lines or symbols that are too much alike may be difficult to follow.

8. Be consistent with abbreviations and other symbols from slide to slide in the same presentation. This consistency should include color, symbols, axes, or any other communication device that is intended to carry the same message in one slide as in another.

9. Avoid the use of too many different colors. Any device for enhancement (italics, bold print, underlining) that attracts attention to itself but serves no additional purpose is a distraction from the content.

10. Use animation sparingly but appropriately. Use clip art rarely; your own photographs will serve you better.

11. Ask a colleague to review your slides and offer an opinion. What appeals to you may not appeal to someone else. You may want a second or third opinion, but if a point is questionable, there is probably a better way to compose the slide.

12. Make your slide set long before you need it, and be willing to remake it.

SLIDE PRODUCTION

A number of techniques are available for making word slides. One method is not necessarily better than another; success depends, first and foremost, on your slide composition. No technique or film will make a good slide from poorly composed copy. Beyond the copy composition, the effectiveness of your slides depends on your judgment and your skill in using a chosen process. If you don't have access to a computer, quality slides can still be made with camera and film. Study the literature from the 1980s on slide making. A 35-mm camera with a macro (50-mm) lens and a light stand can give you good close-up shots for text. However, computers have made slide making much easier. Film recorders attached to computers produce the images electronically from software, and some very good software programs are available. Try to choose a program and film recorder that can provide gradients in type sizes as well as several type styles, colors, symbols, and other embellishments that will allow you to follow the principles of good slide output. Of course, if you have an electronic projector for your presentation and software such as Microsoft PowerPoint® to design your slides, you have advantages of being able to compose and revise slides quickly. This technique is certainly the one recommended if the equipment is available.

No matter what method you use, keep in mind the standards for slide composition. Even though the computer can produce multiple boxes, three-dimensional images, thousands of colors, shadows, animation, and intricate details, **a simple slide is still the best slide** (see Appendix 13). Animation or computer clip art can help or hurt your slide. A neat unobtrusive logo or a new point moved into the slide and a previous one faded to the background with animation can help audience understand, but don't get too carried away with the art work; it can clutter up a perfectly good slide. You want your audience to focus on the scientific message, comprehend that idea in a few seconds, and keep their attention on what you are saying. Don't let that focus be distracted by something clever that your computer program will do.

Slides and other visual aids can contribute to the success of an oral presentation. **But it is better to have no visual aid than a bad visual aid.** Computer technology has not eliminated poor judgment on the part of speakers. At almost every professional meeting I've attended, I have witnessed some speaker who projected a totally illegible slide on the screen and told his or her audience, "I'm sorry this slide is bad, but here you see...." Then the speaker would proceed to point to something I could **not** see. Don't apologize for a bad visual aid; don't use it!

References

Anholt, R. R. H. (1994). *Dazzle 'em with Style: The Art of Oral Scientific Presentation*. W. H. Freeman & Co., New York.

Plotnik, A. (1982). *The Elements of Editing: A Modern Guide for Editors and Journalists*, pp. 121–143. MacMillan, New York.

Smith, T. C. (1984). *Making Successful Presentations: A Self-teaching Guide*. John Wiley & Sons, New York.

16

THE ORAL PRESENTATION

"To speak much is one thing, to speak well another."

SOPHOCLES

You have done scientific research on a certain subject and have established a degree of expertise in that area. You are ready to make an oral presentation in your departmental seminar, at a professional meeting, at a job interview, or to a local civic club. My first suggestion is that you read "Let There Be Stoning" by Jay Lehr (1985). A copy of that editorial appears in Appendix 14. Lehr suggests that boring speakers waste our time and that they should be punished. We should demand excellence in professional presentations the same as we do in published papers. After reading Lehr's vivid analysis of what it takes to make a good slide presentation and before you start designing slides, think about who makes up your audience and what you want to tell them. With the **audience** foremost in your mind, consider your purpose and subject, your own personality and ability, the time you have, and any other influence on the outcome of your presentation such as the physical setting, other speeches, and the presence or absence of a moderator. Then prepare your presentation.

You may wish to write a draft of a speech, but don't memorize it or read it to your audience. Many people can work better from a detailed outline than from a fully written speech, but don't rely totally on the written word in any form. Let your slides or a brief outline with key words guide you. Make the speech in conversational tones with as much eye contact with the audience and as little reference to notes as possible. Most importantly, condition yourself, allow time, construct a slide set you are proud of, organize the material carefully, and practice with a reviewer.

CONDITIONING YOURSELF

Before any oral presentation, you need to prepare yourself for the intensity you should feel while you are speaking (Fig. 16-1). If you are a good speaker, you will feel this intensity, so don't let it frighten you. You need to practice, but you also need to be relaxed and enthusiastic about your speech. If you fail to get enough sleep or if you practice repeatedly up to the last hour before the talk until the speech is almost memorized, you may sound tired and bored with the entire presentation. It is far better to prepare the presentation several days ahead of time, have someone review it, revise it, and then put it aside until just before you go on stage. Get a good night's sleep, dress appropriately for the event, and approach the talk with confidence and enthusiasm.

If you have questions about the organization of your report or if there are technical points that are not clear to you, spend some time before the presentation talking with your major professor or other colleagues who can help you. Be sure you are familiar with the background research (literature) that has been done on your subject and that you understand the scientific principles involved. If you do not understand your subject, you cannot expect your audience to understand it after listening to you.

Often, it is important to publish or hand out copies of an abstract before you begin your presentation. Many organizations publish abstracts to all presentations well before their meetings. A well-written abstract can be a great advantage because the audience is listening for what you have to say about the major points

FIGURE 16-1
Prepare yourself for the intensity you will feel as you begin your speech.

in that text. An abstract must be very short or its purpose will be defeated. If the abstract is a handout for the meeting, a list of pertinent references can accompany it, but the abstract itself should contain no literature citations.

The abstract is an impersonal introduction to your subject. A moderator may give a more personal introduction. Then the floor is yours for the designated time. If you are nervous before and during your presentation, do not be unduly disturbed. Most people feel some nervousness in the speaking situation; use that feeling of intensity to make you alert, eager, and animated. Confidence will soon take over if you are well prepared. Above all, don't let the nervous energy make you talk too fast to be understood.

Try to enunciate clearly. Make a special effort to speak slowly and loud enough so that those in the back of the room can understand you. Face your audience as much as possible. Look at them; eye contact is important for holding their attention. Try to avoid such nervous mannerisms as hiding your hands in your pockets, fiddling with objects in your hands, or unnecessary waving of a pointer; these displays are extremely distracting to the audience.

The pointer, the remote control for slide advance, and a microphone are equipment you might use while you are speaking. Practice with various kinds of these tools. You may be issued a light pointer or a stick pointer. Learn to be comfortable with either. A pointer can be helpful to direct and hold the audience's attention to a point on the screen, but turn it off or lay it aside when you are not using it. Otherwise, the light pointer can turn into a distracting series of flashes across the audience, and the stick can appear to be a baton and you an orchestra conductor. The light pointer may be on the remote control, and that device may also contain signals for turning the projector on and off, for focusing, and for advancing to the next slide or returning to the previous. The equipment in a meeting room may not be identical to that you practice with. To avoid appearing incompetent, become acquainted with it, if possible, before you are introduced. Pushing the wrong button can disrupt your talk.

The microphone is a valuable accessory especially in a large room or for a voice that does not project well. Take advantage of a good mike; it can help you provide quality in tone and enthusiasm. As with the pointer, you need to become comfortable with the mike. You may have to adjust to the kind used, and more problems occur for women than for men. The mike may be a clip-on that can be attached easily to a tie or suit lapel, but it can be unduly heavy on a thin blouse or dress and even worse if the garment has no lapel. The microphone may be stationary on the podium, or it may be on a cord that is put around your neck. These can be better for women, but anyone needs to be conscious of the range of the stationary one and not walk too far away from it. You may even have a mike that uses a pocket remote sensor to a speaker system. These are simple devices that slip easily into the pocket of a man's coat or pants but may not be suitable for some women's clothing. As microphone systems become perfected and manufacturers recognize the need, perhaps women will not have to adjust their attire to fit the equipment.

TIMING

Recognize that timing is important, not just for beginning and ending a speech, but also during preparation and delivery of any talk. Boil down your remarks so that you can present the significant points and auxiliary explanation within the allotted time and give adequate coverage to each point of emphasis. As McCown (1981) suggests with posters, a good idea is to start with your abstract and enlarge on the points it contains with examples and details. Planning for a speech and designing visual aids should begin at least 3 or 4 weeks before your presentation. In fact, as you do your research and data analysis, you should be planning for the delivery of results to an audience. Take photographs and design other visual aids throughout your study. Even with modern technology, you should plan to have your presentation in its final form at least a week before the actual delivery. In addition to conditioning yourself to make a fresh and enthusiastic presentation, this time allows for the possibility of correcting errors in slides or improving poor communication efforts after a review by your peers.

You will seldom have an opportunity to set your own time limit for a presentation. For some situations, such as a job interview seminar, you may be given a generous range of time. A prospective employer may test your judgment by suggesting that you take 20 or 30 minutes, but no more than 40. Wisely plan that presentation at 25 minutes; your audience will probably want to ask questions, and the attention span for most people is less than 30 minutes. In other situations you will be given a designated time within a program. That time may be very specific: 15 minutes, use 12 minutes for the speech and 3 minutes for questions. Such a request gives you very little leeway, but you will note that others have the same limits, and it is rude and unprofessional to finish too soon or to run over into someone else's time. The program chair expects that you will fill your time slot and does not want to get off schedule by starting the next presentation too early or too late.

Timing is also critical during the speech itself. Give your main point the most time, and don't spend much time on side issues. Usually the time allotted will allow you to make only one main point or possibly two. The way you use the time for exhibiting those points depends on your organization. The pace with which you move through the presentation is critical to audience reception. Anholt (1994) suggests that "Slowing down is a remedy for 90% of most speakers' problems." Slow down and remind yourself to provide the audience with time to think by using pauses, full stops, and a relatively slow pace. A friend of mine, who is an outstanding speaker, writes **slow down** as a reminder at the top of each note card. If you have a tendency to talk fast, you might do the same. The speaker who gives the audience a second or two to think just after he or she makes an important point usually maintains the audience's attention far better than does a speaker who drones on without clear punctuation in the delivery.

THE VISUAL AIDS

Even with the same subject, visual images should be somewhat different for any given audience. Your colleagues in science may understand a complex graph that would be lost on the audience at the job interview or the civic club. The latter audience might enjoy seeing a picture of equipment you have used, but the scientists may have seen such instruments so often that if you simply name the piece of equipment, an image is in their minds without the need for a photo. After you have thought about your audience and your purpose, only then are you ready to consider the content for slides.

Seldom should you put words on a screen and then say those exact words without further comment. In other words, don't read your slides to an audience. Exceptions to this idea can be your title, your objectives, and your conclusions. The emphasis with visual and verbal duplication is not only acceptable but can be highly effective. Be selective in what you duplicate. You may not wish to read the title verbatim, but it is usually best to read the objectives and conclusions and add commentary with each. For almost every other part of your speech, do not bore your audience by putting the text of your speech on the screen. Instead, select key words and other images, display them, and expand on their meanings in your talk. This technique allows you to keep text on the slides that the audience can comprehend at a glance and then move their eyes back to you to get the most from your voice and body language. Use a pointer discreetly to direct or hold their attention to a point on a slide, but lay that pointer aside when you are not using it. Study the principles for creating and using any visual aids before making your slides (see Chapter 15).

COORDINATING THE VISUAL AIDS AND THE SPEECH

As you organize your talk, consider the audience and the time you can give to each portion of the content. Speech teachers may tell you that you should roughly divide your talk into 10% introduction, 80% main body of the speech, and 10% conclusion. Often these percentages should be altered to provide more time at the beginning or the end for explanations that are crucial to a particular audience's understanding. In scientific communications, we may follow the classical

TABLE 16-1

Typical Content for a Presentation Reporting Scientific Results

Speech Content	Slide Content[a]
I. Title	Title and authors
II. Introduction	
Hypothesis	Full statement
Justification	Key words, pictures
Literature	Ideas and references
Objectives	Full statements
III. Methods	
Equipment	Photos or illustrations
Sampling and technique	Lists or flow charts
Methods of analysis	Statement or key words
IV. Results and Discussion	
Objectives accomplished	Statement and pictures
Data	Tables, figures, key words, photographs
V. Conclusions	
Main outcomes	Full statements
Accuracy of hypothesis	Statement or photograph
Application of results	List or picture

[a]These are examples that usually work well. Your content may require that you use other forms.

beginning, middle, and end, but in research reports we are more likely to think of organization in terms of introduction, methods, results, and discussion. Table 16-1 outlines typical contents for both the text and the slides for a research presentation.

For any speech, with or without visuals, your introduction needs to let the audience know what your subject is and how you will approach it. Once you have identified your subject and provided a justification or hypothesis and the specific objectives for the substance of your talk, the best introduction is usually a general outline of what will follow: "Today, I'd like to discuss two issues.... The first is..., and the second...," or "We can follow the development of this theory through three stages: 1..., 2..., 3...." The audience can follow you more easily if they have that initial outline of information. From that point, take the speech through the development of your points, and draw a conclusion by reiterating the main points and suggesting an interpretation based on your development of ideas. With any speech, decide what you want the audience to know and carefully set up your objectives and conclusions with adequate intervening material to support both. Then ask for a second opinion about how well you are communicating and revise the order if need be.

Paramount to making your slides aid you in your presentation is to have them well organized and carefully coordinated with your speech. **A title slide, an objective slide, and a conclusion slide** form the foundations around which you organize your talk and your other slides. Even these may be modified for different audiences, but they will exist in one form or another in any good slide presentation.

The **title slide** should focus on what you are going to talk about. It should be clear and aesthetically pleasing. Beyond your own appearance, it may be the first impression you make on your audience. In addition to the title itself, this slide should contain your name and the institution or agency you represent (if you do). Some people argue that, on most occasions, a host or a moderator will have introduced the speaker by name along with the title of the presentation, and therefore, the slide need not repeat that information. I believe such information bears repeating. Here is another legitimate use of duplication in speech. Even though your title and your name have been announced, a visual aid can reinforce that information or provide it to those who did not hear or understand the moderator. Your title and identity are important; double the listeners' chances of having that information.

The **objective slides** should briefly state the research objectives that you are going to discuss or should show the audience the purpose or outline of the speech you are presenting. Objectives should be limited in number; it is hardly feasible, even in a relatively long speech, to present more than three research objectives, and one or two are better. They should be worded as briefly as possible, but don't neglect clarity for brevity. You may wish to read each objective verbatim from the slide and then pause and discuss its meaning or why you have chosen that objective. Move slowly over the objectives; you may want to put each on a separate slide to slow you down or use animation to focus on each objective individually. The objectives constitute the heart of your discourse; be sure to focus on them and give them the time and emphasis they deserve.

The importance of the **conclusion slides** also cannot be overrated. The listeners may forget most of what you have said, but if you will present dynamic conclusions, they will be able to carry your point away with them. Remember that the content in your conclusions should be coordinated with that of the objectives. Your objectives should state your goals and the conclusions should indicate the extent to which you accomplished them. On the slides, conclusions can be in complete sentences, and as with the objectives, you need to move through them slowly and deliberately. Embellish each with reminders of what you have said in the body of the talk that led to the conclusion. Above all, keep your voice alive and enthusiastic. If someone's attention has wandered, it is time to demand that your point be heard. Use pleasant but forceful words and voice; punctuate with appropriate facial expressions and hand or body movements.

Once you have designed your title, objective, and conclusion slides, you are ready to fill in between them. Your first task after with the title is to make clear your point of emphasis, your subject, your hypothesis, and the justification for doing your study or for presenting your ideas. Even before your title slide, you may wish to show a photograph or two that provide justification for your study or a visual image of a species or a condition you will be discussing. For example, if you have been studying the habits of the American burying beetle, you may want to show a close-up of the beetle itself. Not many people have seen one in recent years. During the showing of two or three such photographs (no more than that), you can explain the value of the beetle in the environment, the fact that it is now endangered, and the need to understand its habitats and habits. Some of that justification can be done before the title or all of it after the title, but along with the objectives you will want to enlarge on the justification as much as your introductory time will allow. Present your own hypothesis and reference to literature that relates directly to your study. For instance, you may need to compare or contrast your work concerning the American burying beetle with that on similar species. Objectives may conclude the introduction after you have justified your study, or they may come soon after the title prior to that justification. No one pattern for organization is better than another, except for a given subject and a given speaker. In other words, the **introduction** of your speech and the visual aids that accompany it will include a title and objectives with other supporting material that can be arranged in any of various patterns around these two essentials.

The remainder of your presentation will be much more lucid if this introduction is clear. Characteristically, in a presentation on your research findings, you will move from the introduction to the **methods**. Think of what unspoken questions may be in the minds in your audience. They may be pertinent, or they could get you off track. Stick closely to your purpose. Your responsibility is to answer the right questions and guide the audience to your point. If you have been studying the American burying beetle, your audience may want to know how you trap the beetle, how you get a license to handle the endangered species, what data you collected, under what circumstances you observed the beetle, or what analysis you used; the list of questions can become endless. You must decide which questions are most relevant to the focus of your speech and how to answer them in the time you have to devote to methods. Let your visual aids then support your answers. You may have slides in the form of diagrams of the traps you use, as well as photographs of the area where you trap beetles or of the traps themselves. You may have lists of materials, a statistical design, or categories of data such as eating habits or flight habits of the beetles. For an audience that is mostly interested in your results, do not spend too much time on methods unless your objective was to study methods. Your basic design for the experiment, the kinds of data collected, and the statistical analyses used usually provide enough information to tell an audience that you have pursued your objectives in a scientific manner; then go on to the results.

Again, your audience is crucial to the way you handle the **results and discussion** section. Present representative, selected data to illustrate what you have accomplished in pursuing your objectives. Discuss any special meanings and implications that can be drawn from the study. Fitting your study in with those of others researchers further increases your credibility. Make reference to specific literature and use the names of other researchers. Such specific references will enhance your reputation far more than a general allusion to "other researchers."

For those at a professional meeting who understand your technical and scientific terminology and are accustomed to seeing line graphs and bar charts as well as tables of information, you can rely heavily on these forms to present results. For another audience, you may need more word slides or photographs. Certainly, for any audience a mix of word slides with field and laboratory photographs and graphic depiction of data is more informative and pleasant than is all words or all photographs. Notice the mix in Appendix 13. Precision and clarity in communication can be increased by using a variety of different techniques for presenting an important point. In your presentation on American burying beetles, you might show results by providing numbers in a table and a bar chart of information, and then to illustrate your observations and discussion pictures of the beetle itself.

Hold to your point of emphasis throughout your presentation. As with the methods, results and discussion should flow from the objectives to establish whatever resolution your data provide for your hypothesis. Remind the audience of your objectives, and then show the results and discuss their significance relative to those objectives. Once you have clearly enumerated results and discussed them in relation to your main point of emphasis, you are ready for the **conclusions.** Be sure the conclusions clearly reflect the goals of your objectives.

Throughout the presentation, carefully coordinate the slide on the screen with the words you are speaking. Practice enough to be sure that the order of the slides is the order of the talk and that you are comfortable with both. Critical to a smooth flow of the speech from beginning to end are the transitions that punctuate the presentation.

TRANSITIONS IN A SLIDE PRESENTATION

Often the difference in an adequate presentation and an excellent one results from the pace of the speaker and the transitions he or she uses to move the audience along. Transitions slow you down and give the audience time to think. Transitions get you into and out of the speech and move your content from one section to the next, from one point to another, and even from one slide to the next. The typical slide presentation based on a research project contains several strategic junctures at which you must carry the audience from one point to the next with clear transition. Internal transitional devices such as those used in written work (see Chapter 3) can be used, but the situation in speaking is different, and the speaker must guide the audience with transitions through this unique format.

The first transition many people overlook in oral presentations is the juncture that occurs as you begin the talk. If you have no one to introduce you, you will have to attract the audience's attention to you and your speech. Some in that audience are talking among themselves about subjects entirely unrelated to your speech, some are thinking about their schedules or their own speeches, and some are simply concerned with the weather or their comfort. Transitions here include such things as closing a door, stepping to the front of the audience, stepping forward toward the audience, dimming the lights, or saying something such as, "Let me have your attention, please." Most professionals interpret all of these gestures as an indication that you are about to begin. You may have to pause as the talkers finish a sentence or a latecomer finds a seat. The important point is to have as much of the audience's attention as possible directed toward you and what you are saying.

Of all the transitional devices in the world, the *pause* is often the most effective in speaking. After you have made a remark, closed a door, and stepped forward toward the front row of your audience, a pause will do wonders to bring the audience to you. Even those in their own private worlds of thought will look up at a silence to ask themselves, "What's going on?" or "What's the speaker waiting for?" The key is to pause long enough to attract the attention but not long enough to lose it again.

A part of the responsibility for moving the attention of the audience to the speaker may belong to a moderator. This host may be the one to shut the door, adjust the lights, call for attention, and make the strategic pause. Whether you or the moderator assumes these duties, the transition is not complete until you begin your speech. Once you have placed yourself in front of the audience and are ready to begin your speech, give the audience a moment to adjust to the way you look and sound. A smile, eye contact across the audience, and an introduction to your voice are in order. Thank the one who introduced you, and then let your first sentence lead the audience toward your point, but do not let it be critical to their understanding your main point. Their minds have a lot of baggage to check before they can join you. They have to put aside whatever else they were thinking and get used the appearance and sound of you. Making this initial transition will probably take no more than 10 seconds, but it is crucial to a good beginning.

Equally important is the manner in which you leave the audience. Without a transition, an awkward moment occurs as the last word in your prepared speech is spoken and you are ready for questions or to leave the audience. A moderator can be valuable at this juncture. If the host will ask for questions, readjust the lights, or dismiss the audience, you have the moment to let your mind relax, turn off your last slide, or prepare to answer questions. If you are alone, you must assume the responsibilities of both you and the moderator. At this final juncture, you must recognize how much time you have left, maintain a professional stance, and let the audience know what comes next. Do not loosen your tie or roll up your sleeves or stuff your hands in your pockets. If a question period is to follow, do not remove a microphone that has been attached to your lapel or around your neck. Be sure the last slide is turned off and the lights are turned on for the audience. Whether you or the moderator asks for questions, a good transitional gesture is to move a step or two toward the audience. This symbolic language suggests an invitation for the audience to join you; you are joining them. After you have answered the last question, the moderator may assume the transitional duties of making a closing remark, thanking you and thanking the audience for coming, and providing any additional instructions (e.g., there will be another speech, the management has asked that we exit to the right). Moderators are handy things to have. Without one, you need to summarize the talk and the questions, provide any instructions the audience needs, and thank them for listening to you. Above all be sure to finish in the time allotted to you.

Perhaps you had not thought of the periods of opening and closing the speech as transitions. Treat these entrances and exits with respect. They may be the most important transitions you make. More likely you think of moving from one section of your speech to another as needing transition. How do you leave the introduction and move into the methods, or leave the discussion and move to the conclusions? These certainly are critical points at which you must carry the audience from thinking of one thing to thinking of another. Give the audience a smooth ride and they won't notice the transition. For an oral presentation, some speakers will use headnotes on slides to indicate a new section of the speech — Introduction, Methods, Results, Conclusions — these signposts provide obvious directions for your listeners and are often appropriate and inoffensive. However, they can be awkward impositions that could be avoided by using a relevant photograph or speech alone. With or without the headnotes, your words need to carry the audience to your next section. To get into methods from the introduction, you may remark, "To carry out this study, we set up two experiments. The first...," or you may say, "To accomplish my objectives, I first acquired...." These remarks are smoother than a blunt announcement of Materials and Methods and are just as effective. Similarly, as you move into the results and discussion, you may have some sort of transitional slide, but you should also let your words carry the transition. Try something such as, "The data collected showed that our hypothesis was accurate" and then move to a results slide that supports that contention, or you may need to say, "Our results are inconclusive, but they do show that...."

Moving to the conclusions is critical. You must have the full attention of the audience in order to leave them with your main points. Sum up the essence of the results you have presented and move dynamically to the conclusions. "All of these results point to two conclusions" or "Although the data are limited, we can conclude that..." are simple and often effective entrances to the conclusions, but you may be able to be more creative than these remarks are.

Let's go back to the American burying beetle. Your hypothesis may have been that the lights of the homes and cities across the country are attracting these beetles away from their natural habitats in dead carcasses and from their habit of setting up homes in these habitats to produce and feed their offspring. Your conclusions need to lead us back to that hypothesis and your objectives. You could move from your results to the conclusions by showing a photograph of the beetle near a light or moving toward one and declaring that "Our results indicate that American burying beetles will not keep their minds on home and family when they see the lights up town. We conclude that...." You are ready at that point to exhibit your first conclusion slide, which will most likely reflect the outcome of your first objective. Your remark has given the audience time to digest their last thoughts about the data you have just presented and to move with you to the conclusions. Remember the value of the pause. A deliberate pause with a transitional statement can effectively attract the audience's attention. Let the slide on the screen fit the occasion and serve to support the pause and the remark.

In addition to transitions into and out of the speech and at major junctures, you need clear transitions all along the way. Transitional word slides or photographs can help. Whether a picture or words are on the screen, the important point is that the transitional slide is appropriate. Smooth movement from one idea to another may not allow time for an extra slide. As you linger momentarily over one slide, you draw a conclusion to that point and then move with a short transitional word or phrase to the slide displaying your next idea. For example, as you leave one data slide, you could be saying, "Contrary to what we found with these data..., [advance to next slide] the analysis of...showed that...." Such transition should be woven throughout the presentation.

Be careful not to keep repeating weak transitional words or phrases that become worn thin with over use. Some to use with great caution include the following: "Also, we see...," "And here you can see...," "Again...," "Looking at the..., we see that...." Also avoid nontransitional utterances such as "Uhh," "And uh," or "OK." As Booth (1993) suggests, "'Anderm' [is] the most irritating nonword ever misfangled." Make your transitions provide the audience with both meaning and think time. Instead of saying "again," give the audience a little time to move from one point to another with a transition such as, "Those data show us the anatomical differences, but other data [advance slide] illustrate that physiological differences are just as important." Yes, it takes longer to make that statement than to say "again," but the audience has time to think about anatomy and be ready for the next point on physiology. This think time is essential to any good communication.

THE PEER REVIEW

When you believe you have your content organized, laced together with transitions, and proportioned relative to time, go through your speech with your colleagues. Always accept healthy criticism. Their vantage point is closer to that of the audience than yours is. If slides or speech content and organization are not clear to these reviewers, revise your materials until they are. As you make more and more talks, you will become more sensitive to what works for you and the audience. When you have gained experience, don't be afraid to go it on your own, but always welcome reviews. In addition, during the speech be sensitive to the audience reception and be willing to deviate from your prepared material if you recognize that their attention or understanding is straying.

These general comments on timing, organization, and review will not make a good speaker of you. As with learning to swim, the only way to develop the skill is to do it. If you follow the basic conventions for good speech making and repeatedly practice the proper moves, you can become an accomplished speaker. Anholt (1994), Booth (1993), and Appendix 14 (Lehr, 1985) have good suggestions for making scientific presentations. The following check list may alert you to points to observe in your own scientific presentations and those of others.

CHECK LIST FOR SCIENTIFIC SLIDE PRESENTATIONS

I. The Speech
  A. Introduction
    1. Are your hypothesis and objectives clear for the audience?
    2. Do you provide the audience with clear rationale and justification for your study?
    3. Does your introduction follow a logical pattern, and is it related to other literature and to scientific principles?

B. Materials and Methods

1. Do your methods have the support of the literature and scientific principles?
2. Do you show a logical, step-by-step process for executing the experiment and collecting the data to carry out your objectives?
3. Do you make clear your use of appropriate experimental design and statistical analyses?

C. Results and Discussion

1. Do you summarize results, that is, emphasize main points, as you **begin** and **end** this section?
2. Do you relate the results clearly to your objectives?
3. Do you carefully choose a limited number of data points to support your contentions and present them in simple illustrations, graphs, tables, and lists?
4. Do you discuss your points in terms of:
   a. Their relation to other research?
   b. Their practical or scientific applications?

D. Conclusions

1. Do the conclusions reiterate main points for the audience to remember?
2. Do you show a list and clearly relate it to your objectives?
3. Do you give examples of application and use for your findings?

II. Visual Aids

A. Number — Are there too many or too few slides for the time you have?

B. Content

1. Are the slides clearly coordinated with your speech?
2. Is the purpose of each slide readily apparent?
3. Do you have a balance of data, lists, and information slides with photographs interspersed throughout?
4. Have you included all expected slides: title, list of objectives, conclusions?

C. Quality — Are your slides:

1. Neat and spaced to fill the screen?
2. Simple and free from excessive data?
3. Easy to comprehend, that is, have proper size print, good content, good design, clearly labeled axes?
4. Free from garish color or any other embellishment that could distract from your message?

III. Speaker

A. Are you prepared?

1. Are you familiar with your speech and the slides?
2. Will you and your audience be comfortable with your appearance?

B. To what extent do the following support or distract:
1. Mannerisms and gestures?
2. Audience contact (eye contact, facial expressions)?
3. Voice, speech patterns, and ease in speaking?
4. Your attire, posture, and poise?

C. During the speech, keep the following in mind:
1. Avoid reading from the slides or from notes.
2. Be sure your eye contact covers all the audience.
3. Put the pointer down or turn it off when you're not using it.
4. Don't put your hands in your pockets.
5. Never apologize or make excuses for a bad slide. A bad slide is worse than no slide.
6. Keep your voice enthusiastic and loud and slow enough.
7. Be sure to use enough but not too much time.

Recognize all the elements that come together in a successful slide presentation, practice them, and be proud of your performance. A good slide presentation coordinates communication by the speaker, the speech, and the visual aids. It has a specific purpose and is directed toward a specific audience. Be conscious of all the influences on the speaker, the speech, and the visual aids. Check everything, including the physical setting, the pointer, and the microphone. Prepare with good slides, clear organization with appropriate transitions, and a peer review of the presentation. During the talk, be alert to the reception by the audience and the role of the moderator. Maintain a professional attitude before and after the talk as well as throughout the presentation, which includes the question/answer session. Confidence and a dash of humility will lead you to success.

References

Anholt, R. R. H. (1994). *Dazzle 'em with Style: The Art of Oral Scientific Presentation.* W. H. Freeman & Co., New York.

Booth, V. (1993). *Communicating in Science: Writing a Scientific Paper and Speaking at Scientific Meetings.* 2nd ed. Cambridge University Press, Cambridge.

Lehr, J. H. (1985). Let there be stoning. *Ground Water* **23**, 162–165.

McCown, B. H. (1981). Guidelines for the preparation and presentation of posters at scientific meetings. *HortScience* **16**, 146–147.

17

POSTER PRESENTATIONS

"Only the composition as a whole determines the good
or bad of a piece of graphic work."

EDUARD IMHOF

Posters have become a major format for communicating at scientific meetings. The technique varies with different societies, but generally the poster will be on display for several hours, perhaps all day, and the authors will be present during a part of that time to discuss the subject with viewers. Depending on the meeting, the number of posters displayed at one time may range from a dozen to several hundred. At any one time, the audience for each poster is a relatively small group of interested people. Presenting a poster is a good opportunity to build your reputation as a confident, knowledgeable, articulate scientist if you exhibit an attractive, informative display and maintain a professional demeanor as the author.

Posters were practically unheard of 35 years ago. They were introduced into scientific meetings in the United States in the mid-1970s (Maugh, 1974). They rapidly became a way to display large numbers of research reports and have been widely accepted as a viable complement and alternative to slide presentations and symposia or workshops. Posters offer advantages both for meeting arrangements and for communication efficiency. More posters can be scheduled in less space than can oral presentations, and those attending meetings have access to more research in the same amount of time. Some convention centers can now provide large areas and display boards more easily than they can provide numerous meeting rooms and visual projection equipment.

The advantages to individual communication are as appealing as the tactical convenience in arranging for a convention. The method provides a two-way interaction that is less feasible with the slide presentation. The poster presentation may turn into a very profitable question and answer session, with both the presenter and the audience deriving mutual benefit from the ideas exchanged. Compared with the slide presentation, the poster technique provides more convenience in following up on ideas. Names and addresses or phone numbers can be easily exchanged, and the scientist will be more likely to contact the person he or she has spoken with face to face.

The poster is obviously a good format for scientific communication; however, it has evolved so rapidly, compared with a few hundred years for the evolution of the published or spoken scientific report, that we still struggle with authoritative guidelines relative to form, content, and construction of posters. Imhof (1982) declared that "Only the composition as a whole determines the good or bad of a piece of graphic work." What is true for Imhof's maps is also true for posters. A successful poster must communicate through every visual and verbal detail. Because posters are essentially a cross between the visual and spoken communication used in a slide presentation plus the textual substance of a written paper, the same communication devices can be adapted to the poster format. Posters perhaps balance the visual, oral, and written elements more fully than either the slide presentation or the written report. In all three forms, the basic communication devices include the text, the type size and style, color and texture, shape and arrangement, and illustrations of data in tables, figures, or photographs. In getting involved with all of these media, keep in mind the concept of unity or "the composition as a whole," and remind yourself of the basic purpose of any scientific communication: to convey clearly a scientific message to an interested audience.

AUDIENCE

As with any other communication, with posters you need to have as much concern for the audience as you do for your subject and your own presentation of it. These viewers are 1 to 2 m away as they read the poster (Fig 17-1). The material must be attractive, interesting, and clearly legible to keep their attention. Meetings offer many possibilities, and your viewers have plenty of other things to do. Also, most readers are standing, and it is tiring to read from the standing position for long. Most of them will look at the main points of your poster and then move on. O'Connor (1991) says that a typical poster reader will stop, read, and move on—all in 90 seconds or less. You are competing, then, with other activities at the meetings, with tired feet and eyes, and with viewer's time. However, the viewers are there because they are interested in your poster and your subject.

FIGURE 17-1
Your audience is a small group of truly interested people.

Design and present your poster in such a way that you keep their interest and that the experience is worth their time and yours.

Woolsey (1989) suggests that your poster audience can be categorized into three groups: (1) colleagues who follow your work closely, (2) those who work in the same area but not on the same specialty, and (3) those whose work has little or no relationship to yours. He suggests that the middle group is your target audience. Those in the first group will probably stop at your poster to see how you have presented information they already are familiar with and to see if you have come up with any new data or details. Viewers from the third group are not the audience that you need to attract anyway; they simply pause to catch your main point and, not being interested, move on. Members of the second group are interested in your subject and are not familiar with your data. If your poster is clear and attractive, they will probably stop longer than O'Connor's (1991) 90 seconds. To accommodate all three groups, your poster should be:

—**Brief and clearly organized**
—**Simple with an obvious central point**
—**Easy to read from 1 to 2 m away**
—**Attractive and aesthetically pleasing**

A closer look at the details that go into a poster can help you satisfy these criteria.

TEXT

Briscoe (1996) is right when she maintains that "It takes intelligence, even brilliance, to condense and focus information into a clear, simple presentation that will be read and remembered. Ignorance and arrogance are shown in the crowded, complicated, hard-to-read poster." The poster format demands concise presentation of information and clear coordination of words and visuals. The content should be a full but brief report, not a full paper but a synopsis. In contrast to the written paper, the poster may omit the abstract, provide little discussion, and use more photographs and color. Because the poster itself is an "illustrated abstract" (McCown, 1981) and because the abstract is likely printed in proceedings of the meeting, it is not needed on the poster unless the sponsoring society requests that it be posted.

Some viewers may peruse the display when you are not present. The poster should reflect your credibility as an author by providing substantial justification for your objectives and by giving meaning to your results and conclusions, but extensive written discussion is out of place. Visual communication with pictures and other illustrations can be valuable. You should be present at the scheduled time to answer questions and discuss issues, but be sure the text of the poster will stand on its own for those unable to be there at the same time.

As you plan the content for your poster, the following ideas can aid in the audience's understanding.

—**Let nothing distract from your scientific message.**
—**Use short expanses of text and short paragraphs. More than 20 continuous lines will tax the audience's patience.**
—**Present lists when possible especially in such sections as the objectives or conclusions.**
—**Use visual imagery and mix it freely with the text.**

—Select type style and size, colors, and spacing that make reading the text easy and pleasant.
—Provide appropriate handouts such as business cards and a summary or abstract with your name and contact information included.

With these points in mind, carefully consider the content and organization of your poster. You are not writing a paper for publication. A full paper can be more complex because it can be read from a comfortable sitting position and can be studied over a longer time. Writing the poster content into a concise, meaningful text without making it too long is a chief concern for anyone designing a poster. Rather than trying to condense the paper, you should expand the abstract; that is, deliver the concise, precise format of "an illustrated abstract of a publication" (McCown, 1981). To avoid long stretches of unbroken text, break in between ideas or sections with a new heading or a picture, table, figure, or other illustration. Long-running paragraphs are formidable, and few in your audience will read them through. Concise lists should substitute for running text wherever possible. You can expand any discussion orally, and handouts of additional materials can be made available.

Especially discouraging to a reader are long lines and long sections of text. As Woolsey (1989) suggests, each line should be comprised of no more than 65 characters and spaces. A section of text should not be more than 20 lines long, and fewer is better. No matter how concise you make your poster, most of your audience will not read it from the beginning to the end. They may read conclusions or objectives first, or they may start by viewing data in tables and figures. Some may begin with the introduction, but especially if the text is long and hard to read, they will begin to skip around among parts. For this reason, the poster should be carefully organized, and each section should carry your central point of emphasis.

Organization for the poster should follow the same organizational convention used for other scientific writing. See examples in Figures 17.2 and 17.3 and Appendix 15. Let the **Introduction** briefly provide the rationale and background for your study and present the objectives. Be sure that your hypothesis or research question is stated early and that the **Objectives** stand out in a list to themselves with a headnote or bolder print to call attention to them. Keep the **Materials and Methods** brief unless your purpose is to present a new method. Wherever you can, use a picture, flow chart, or list to describe equipment or steps in a process. Just be sure the methods section provides enough information to make your study credible and to make clear how the data were derived.

Once the objectives are understood, **Results** are the most important part of most presentations. Limit the text, but use clear tables and illustrations for the data. Present only enough representative data to make the point in your results.

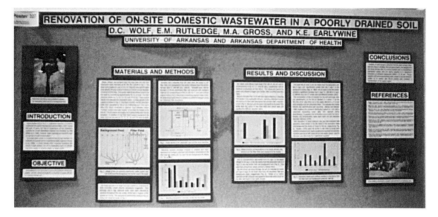

FIGURE 17-2

Spacing is a valuable part of communication with a poster.

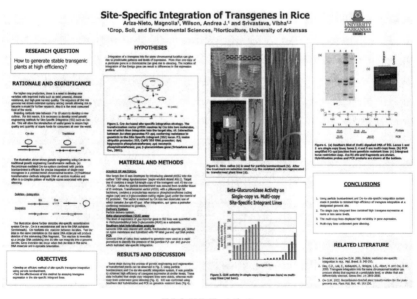

FIGURE 17-3

A well-designed poster can be printed on a single sheet.

Use text as necessary to explain or draw attention to a significant point, but be sure tables and figures are clear without the text. **Discussion** should be limited and is often included with the results under a single heading of **Results and Discussion**.

Emphasize the **Conclusions** under a separate heading and keep them brief and in a list if possible. If readers can quickly find the objectives and conclusions on your poster, then they know whether they want to look into the results and methods sections more thoroughly. If you cite literature in the text, be sure to provide **References;** these can be listed in smaller type and in a less prominent position than other sections of the paper. An **Acknowledgment** is appropriate to recognize contributions to the research or the construction of the poster.

TYPE SIZE AND STYLE

Readability is of primary importance. Just as the message on a slide must be legible from the farthest seat in the room, so must the message on the poster be easy to read from the customary standing position of the viewer. The title needs to be legible from a distance of 5 to 10 m; the text, from 1.5 to 2 m. Readability depends on the size and style of type. Size is measured in points relative to the tall letters in the font. Blocky, thick styles of letters no less than 23 mm for titles and 5 mm for text will accommodate the requirement for legibility (Table 17-1). Common type faces that are good for titles include Arial, Helvetica, Tahoma, or similar ones. For the text most of these are good, or you may choose a conservative serif type such as Times New Roman, Bookman Old Style, or Palatino Linotype. Avoid ornate or script styles, and use italics only where grammatically or scientifically required. A mix of capital and lowercase letters is easier to read than all capitals. As O'Connor (1991) and others suggest, you should use capital letters only where they would be used in conventional texts. The principle of giving the audience what they expect holds true here; our eyes are accustomed to lowercase, with portions of some letters extending below lines and to different heights above the lines.

COLOR AND PHYSICAL QUALITIES

Color, depth, quality, and texture of materials can contribute to the physical appearance and the communication of the poster, or such features can serve as distractions from the scientific message. Refer to Imhof (1982) for principles of communicating with color (Appendix 12). Background paper for the text and the illustrations should be white or a very light color such as beige or parchment; these colors are comfortable for the eyes. Onto this background can be put

TABLE 17-1

Recommended Type Sizes for Posters

Sample	Font Sizes (Height)	
	Shown	Range
Title	120 (30mm)	90–144
Heading	60 (15mm)	30–90
Subheading	30 (8mm)	30–60
Text	24 (6mm)	16–30

All examples are in Arial style.
Originally published by Davis et al. (1992) in the *J. Nat. Resour. Life Sci. Educ.* 21:158.

almost any distinctive hue. For text and tables, the conventional color that will not attract attention to itself is black type. Sometime highlighting a data point on a table with a bright color can call emphasis to a point you wish to make. However, remember that excessive highlighting negates the concept of emphasis. For figures on the light or white background, use distinctive, but not overpowering, colors to represent data especially in bars and lines or sections of pie charts.

Too many colors on one poster are distracting, but color can be used as a point of unity to code portions of the paper. Two different points, objectives, or experiments may carry two different but related colors (for example, blue-grey and blue-green) throughout the poster to code them as separate but related sections of a unified whole. Color is best used when a purpose beyond that of simply attracting attention to itself is evident. Poor or inappropriate use of color is especially noticeable in posters. Too many, too brilliant, too pale, or uncoordinated colors distract from the scientific communication in a poster as sure as a clown suit or gym shorts would distract from the words of a speaker at a scientific meeting. See Chapter 14 for further discussion of color.

Part of the use of color depends on the materials used in construction of the poster. The physical appearance of a poster on a single large sheet of paper that can be printed on some of today's plotters is quite different from that of a poster constructed with each section mounted on mat board or poster paper. Although with know-how of the computer software needed, a single sheet has become the easiest way to construct a poster, the mounting of posters in sections can create a very attractive display (Fig. 17-2). Poster board or matting materials are more durable and more attractive than are thinner construction or poster paper. Colors for matting should be dark subdued colors that follow the same principles as outlined for backgrounds for slides. The matting for the poster in Fig. 17-2 is a dark blue, the paper is 65-lb weight parchment-colored paper, and the text is black. A colored background can be used for the single-sheet poster, but because black text is typically printed on it, keep it light. Most of the background for the poster in Fig. 17-3 is pale beige with the Research Question, Objectives, and Conclusion in white boxes. As with slides, some authors have tried reversing the dark type and light background so that the background is a subdued but very dark color and the text is white or light. This unconventional technique can attract attention, but unless the contrast is entirely clear, it can be difficult to read and simply attracts attention to itself and not to the scientific message. Good judgment relative to color, sizes of individual pieces or sections, spacing, and arrangement of materials is extremely important for both methods of construction.

SPACING AND ARRANGEMENT

Before you attempt to make any layout for your poster, find out the exact dimensions of the display board that you will be using. Then construct and place the poster sheet or pieces so that they do not crowd the edges of the board. Communication can be enhanced by the sizes and shapes of the sections as well as their positioning on the board. Too many small pieces can give a "busy" appearance to the composition, but one large block with all the blank space at the periphery can be equally unattractive. Blank space is important. Woolsey (1989) has said that ideally 50% of the poster should be blank. This blank space can be used effectively to separate parts of the poster and to communicate relationships among the parts.

A good idea is to section the poster into modules based on the organization of the material, and space the sections in no more than four or five blocks or columns on the board. This same concept can be used in designing the poster in matted pieces or as a single sheet of paper (Figs. 17-2 and 17-3). Logically, a section spaced at a greater distance from the neighboring section is less closely related than one placed nearer. For example, in Fig 17-2, the authors use more space between columns labeled Materials and Methods and those labeled Results and Discussion than between columns within those sections. The same can be done in spacing columns in a single-sheet poster.

Woolsey (1989) suggests that "the eye looks for edges." Too many edges or small pieces separated with framing or space tire the eyes. To avoid extra edges, construct the poster in modules or columns and place subheadings or captions for figures and tables in the same frame or box with the accompanying text or illustration. Shapes that are not expected (triangles, jagged edges, or cutouts) attract attention to themselves. As with highlighting, unusual shapes used effectively may contribute to the communication, but overuse or inappropriate use will simply distract from the scientific message. Shape and arrangement of parts as well as size and color can serve to draw attention to points of greatest importance, to subordinate secondary material, and to unify the entire poster. With use of conventional headings (e.g., Objectives, Methods, Conclusions) and with parts of the poster grouped logically and unified with color and spacing, the viewer should be able to follow a logical flow of information.

PRESENTATION OF DATA

Points of unity and the relative importance of parts of the paper also apply to the presentation of data whether it is in tabular or graphic form, and any illustration should be placed near to the textual content with which it is associated. Size and spacing in a table with a limited number of data points can effectively convey a scientific message. Ideally, a table would contain no more than 20 items in the field. Highlighting with color can draw attention to an important column or item in a table, but avoid overuse this attention-getting device. Be sure to follow the basic principles for constructing tables, and consider that they should communicate, or stand alone, without the text or the author having to explain meanings. See Chapter 11 on presenting data.

Similar principles are true for graphs. To emphasize a point, the number of lines or bars in a graph must be limited. No more than three or four lines or six to eight bars are best. The main point that the data carry should be illustrated as clearly and as simply as possible, and any additional supporting data or subpoints should be reserved for the oral discussion or for a more extensive journal manuscript. Color and size in graphs are very important. Too many colors and sizes are confusing, but a highlight of color or a line that is thicker than the others can make a point more immediate. As in other communications, it is important to maintain consistency in labeling from one graph to another. If a point is represented by a color or shape in one graph, the same color or shape should be used to represent that point in other graphs on the same poster. Briscoe (1996) is very helpful in demonstrating design and construction of illustrations, graphs, and photographs.

Photographs will help to make the poster attractive and serve as relief at breaking points in the text, but those used should carry an additional purpose in the communication. An appropriate photograph can help the title to convey the subject of the poster or can illustrate visually what the data mean. Photographs, similar to other illustrations, need to be clear and large enough (at least $5'' \times 8''$ or $6'' \times 10''$) to be immediately comprehensible and make a point relative to the scientific message. Be sure the point you wish to make with the photograph is in focus and obvious. Too many or undersized photographs contribute too many edges for the eyes. A matte finish is probably better than a glossy finish for most exhibition rooms, where lighting is usually bright.

Computer technology has made it possible to use a color photograph as a background for an entire poster printed on a single sheet. On the surface, this idea appears to be a creative one and can attract immediate attention to a poster. However, I have seldom seen the idea used well for two reasons: (1) Attention is attracted to the background rather than to the text and data that carry the scientific message, and (2) this background is almost invariably composed of multiple colors and shades of color that make it difficult to read some text printed on it. Both these drawbacks are serious limitations in good communication. Use such a background only if you can overcome both the reasons it is not used well.

THE PRESENTER

The relative informality of the poster situation should not relieve a scientist of the responsibility for clear communication and a professional attitude. Your knowledge of the subject, your candor in discussing the science with others, and your appearance and attitude are important to the presentation. The professional meeting will offer many distractions for you as well as your audience. It is your responsibility to be with your poster whenever you are scheduled to be; some of your audience will make a point of being there to talk with you. You may also face the distraction of friends and acquaintances stopping by to chat about things other than the poster. Let your poster audience take precedence over the social audience. Quickly make arrangements to have the friendly chat later, and go back to the poster audience.

There may be periods when no one approaches your poster, and you feel that you may as well leave. Not so. Audiences are likely to appear one or two at a time. Masses of people will not flock to your poster. One important attraction of the poster technique is that the audience is limited to only the truly interested. If your poster adheres to the criteria for a good poster and if a few people read most of it and talk with you, you have been successful. Whether the audience is one or a dozen, execute your professional role with clear communication, knowledge, and sincerity.

HANDOUTS

A simple handout can enhance the poster presentation. A business card or other form with the authors' names, addresses, and phone numbers along with an abstract or other condensed facsimile of the poster, a list of pertinent references, an important method, or a table or figure may prove valuable to the viewers after they leave the meeting. Some presenters supply a small, but readable, printout of the entire poster.

MAKING IT FIT

Poster display boards differ in size at different meetings. Some are as large as or larger than 4′ × 8′ (approximately 1.3 × 2.6 m) and some as small as 3′ × 3′, or approximately 1 × 1 m. A 4′ × 6′ (approximately 1.3 × 2.0 m) board is also rather common. Some are mounted on a stand or rest on a table and are oriented with the longest dimension horizontal. Others may have the longest dimension vertically and stand on the floor. Before you begin to construct the poster, you must know the space and orientation you will have for your display. Decide how much material you can display effectively in the space allowed. Keep the focus as close to eye level as possible.

In addition to space limitation, you must recognize that, with dozens of other posters also on display, yours will command a limited amount of viewers' time before they move on to others. The most common communication fallacy that I have observed in posters occurs in the attempt to present too much material in the time and space allotted. This problem manifests itself both in too much text and too many data. Be willing to make one or two points and leave your other information for future papers.

In making the information fit the board, the second most likely fallacy is in poor construction that distracts from the scientific message. Keep the display simple and attractive. With colors, shapes, and illustration or with words themselves, no device should be used that is merely decorative and attracts attention to itself and away from the science. Of course, you want your poster to be aesthetically pleasing, but foremost you want the scientific message to be clear. These two objectives can complement each other. As Imhof (1982) says, "Clarity and beauty are closely related concepts." Standards of communication for any single visual such as a chart, a slide, or a photograph are also the standards for the entire poster composition. The completed poster must be able to stand on its own with all communication devices in place. The materials should be fitting for the poster format, and the amount of content and the number of illustrations should not overload the message. The display should command credibility, communicate only a limited number of points, and be accurate, clear, legible, concise, and aesthetically pleasing.

In Figs. 17-2 and 17-3 and Appendix 15, notice the limited text. Much of the information is presented in illustrations and figures. The objectives and conclusions are set apart to be readily visible, and a prominent title and the headings can be read from a reasonable distance. Each poster is arranged in sections, and the spacing helps to carry the communication. In their form and organization, these posters deserve high ratings as scientific presentations.

TIME AND CONSTRUCTION

In constructing posters, time is the antagonist. As you conduct your research, keep in mind that you may need to construct a poster in the future. Take photographs for prints and isolate the data that you will use. Weeks before your poster presentation, select possible photographs to use, set up tables and figures, and write the text. If you also need to be working at other tasks, allow yourself at least 2 or 3 weeks after the text is ready to put the poster together. Allow time to reconstruct after reviews; produce final copy only after reviews and revisions are complete.

If you are not already familiar with software that will allow you to print your poster on one large sheet, study what that software can do but guide it carefully in what you want. For example, across a wide sheet of paper, you could print a single line of type that is 2 m long. Several lines that long are very difficult to read. Stick with no more than 65 characters and spaces per line. Take time to arrange all the materials on the single sheet in logical and easy to read form; this work may require some rewriting of text that will not fit a column or omitting a picture that you wanted to use. A rush job can produce a poorly worded or poorly arranged poster.

The same is true if you are producing a poster in separately mounted pieces. Construction in this way can take even longer than with the computer construction. You must have text copied to heavy (65-lb) paper, cut margins carefully, and mount the copy onto carefully trimmed mat board with reliable adhesive. All these things take time. You need pieces small enough to transport easily to the meeting, probably in a small portfolio case. The single sheet can be rolled carefully and packed in a tube. For both kinds of posters, you will probably use hook-and-loop tape or push pins to attach the poster to the display board. Be sure to pack any equipment you may need.

Several sources of information are helpful in learning to design and display posters. See Appendix 15 for a sample text and poster. Basic principles in visual communication that are valuable, not only for posters but for other visual displays, can be found in Briscoe (1996). Very good concise information on the poster format is that of Woolsey (1989), and O'Connor (1991) offers good advice. Some brief but helpful information is in Silyn-Roberts (2002), Luellen (2001), and Knisely (2002). Other suggestions are typically available from societies that use the format.

References

Briscoe, M. H. (1996). *Preparing Scientific Illustrations: A Guide to Better Posters, Presentations, and Publications*. Springer, New York.

Davis, M., Davis, K. J., and Wolf, D. C. (1992). Effective communication with poster displays. *J. Nat. Res. Life Sci. Educ.* **21,** 156–160.

Imhof, E. (1982). *Cartographic Relief Presentations*. Walter de Gruyter, New York.

Knisely, K. (2002). *A Student Handbook for Writing in Biology*. W. H. Freeman & Co., Gordonsville, VA.

Luellen, W. R. (2001). *Fine-Tuning Your Writing*. Wise Owl, Madison, WI.

Maugh, T. H. II. (1974). Poster sessions: A new look at scientific meetings. *Science* **184,** 1361.

McCown, B. H. (1981). Guidelines for the preparation and presentation of posters at scientific meetings. *HortScience* **16,** 146–147.

O'Connor, M. (1991). *Writing Successfully in Science*. HarperCollins Academic, London.

Silyn-Roberts, H. (2000). *Writing for Science and Engineering: Papers, Presentations, and Reports*. Butterworth-Heinemann, Oxford.

Woolsey, J. D. (1989). Combating poster fatigue: How to use visual grammar and analysis to effect better visual communications. *Trends Neurosci.* **12,** 325–332.

18

GROUP COMMUNICATIONS

"The true spirit of conversation consists in building on another man's observation, not overturning it."

BULWER LYTTON

Whether you work for a private industry or a public agency or institution, you will find yourself involved with group communications. As with individual efforts, preparation and competent execution are important to provide clear communication and prevent wasted time and effort. Selecting a format as well as preparing for and carrying out communication with a group depends to a large extent on whether it involves an audience beyond the group itself.

GROUP COMMUNICATION WITH NO AUDIENCE

You may be asked to serve with closed groups to make plans for research projects, to decide policy, or to evaluate a fellow employee's progress. These professional duties are usually performed without an audience and in the format of a round-table discussion or a board or committee meeting. The **round-table discussion, or buzz session,** is informal. The purpose may be for brainstorming or for a professional sharing of ideas with no formal agenda and no decisive outcome expected. It can be very beneficial in initiating and sharing ideas.

A **board** or a **committee** is a more formal, small group selected from a larger organization to assume a given responsibility. Participants follow agendas and expect outcomes that are usually reported to the larger body through their leader. **Standing committees** serve for indefinite periods; **ad hoc committees** are set up to deal with one particular job and are disbanded when the job is complete. Definitions begin to overlap with a **task force,** which is a kind of ad hoc committee but is usually made up of experts in a given area whose job may require very detailed investigation and coverage of an issue. A team may also be involved with writing a report, proposal, or other document. In this situation it is essential that each member of the team clearly understand his or her responsibilities and contribute to a successful outcome.

Whatever responsibilities are required and whatever format is used, professional ethics dictates that you actively perform to the best of your individual ability to accomplish the group's goals (Wilson and Hanna, 1993). Sometimes individuals are appointed to serve in such roles as chair, recorder of oral comments, editor of a team writing project, or even a devil's advocate to search out problems. Often a group selects from among themselves the persons who will fill such roles, or a given personality may assume a role of leader or antagonist without being chosen for the position. Once the group is selected, the task is often decision making or problem solving with no audience involved.

Decision making involves alternatives. Defining what the alternatives are may require research and discussion. Making the best possible decision can depend on gathering information and even solving problems along the way. But after discussion of the pros and cons of each alternative, the group should reach a consensus or conduct a vote to make the final decision.

In **problem solving** there may be no obvious alternatives. The best solutions can be reached with a step-by-step procedure. As with the scientific method, we begin with a problem, identify or define it clearly, gather and analyze data, synthesize the new information with what is already known, and evaluate the outcome. Reaching a solution in a group involves coordinating the expertise and opinions of several minds and personalities. The group has the advantage of dividing work loads and contributing multiple expertise but the disadvantage of coordinating diverse opinions. Recognizing these conditions, the group must follow essentially the same steps an individual would follow.

Procedure for group problem solving

1. The problem is clearly defined and objectives are set forth and understood by all members of the group.

2. Members of the group plan their individual and collective actions. They may divide responsibilities for gathering information and offering opinions.

3. As individuals and as a group, they devise a plan of action.

4. They act on the plan and analyze outcomes.

5. They evaluate the results of their actions and determine whether the solution was acceptable.

With the group, as with individual problem solving, a poor solution to the problem can result from inadequate or faulty information. Each step in the process must be fully executed so that the next step can be achieved. Groups sometimes get in a hurry to reach the goal and fail to fully define the problem or carefully plan their actions before they proceed. Without deliberate planning, members of the group will begin to stumble over their lack of understanding or the sequence of actions, and then they must return to the definition of the problem, explain it, and map plans again.

The size of the group and the time given to accomplish a goal are crucial to success. A group of more than 10 or 12 people will have trouble involving all the personalities and all the individual ideas. An odd number of three, five, or seven can usually work well together. Planning contributes to effective use of time, but because of the need for input from several individuals, group activity may take longer than communication by a single individual. However, the combined expertise of the group can make the additional time worthwhile.

If your group is too large, **brainstorming** is one way to make progress. The advantage of any kind of brainstorming is that, at this beginning, ideas need not be criticized but just presented for consideration. The entire group might brainstorm until they reach a consensus on what ideas to discuss, but if many personalities are involved, the less-assertive members of the group may be intimidated. The large group can divide into smaller buzz groups for preliminary discussions and then present their ideas and objectives to the full group. Another brainstorming technique for handling a group that is too large is to ask the entire group to be silent for a few minutes while each participant writes on a card or slip of paper what he or she believes are the most important ideas to consider. Better yet, have each member come to the meeting with this information already prepared. Then a recorder or leader will present all these ideas to the entire group. From the list, ideas can be combined and organized and priorities can be established. When specific topics or objectives have been filtered from the random ideas that evolve during brainstorming, then the critical process begins.

Positive progress can be made only with thorough, critical analysis by individuals. As a member of the group, you should feel responsible for researching whatever information is pertinent to the decision making, the problem solving, or the presentation of information. You shirk an ethical responsibility if you come unprepared and expect to know enough already or to let others make the decision. When individuals do not assume individual responsibility for preparation, information, and participation, the group interaction can often degenerate to confusion or, perhaps worse, to a point at which individuals are not thinking for themselves (Fig. 18-1).

FIGURE 18-1
Groupthink involving too many passive minds does not provide the best solution to a problem.

Janis (1982) coined the term **groupthink** to apply to a mentality of group agreement without the individual minds sorting out the best decision or solution to a problem: "Groupthink occurs when a group strives to minimize conflict and reach a consensus without critically testing, analyzing, and evaluating ideas" (Beebe and Masterson, 1994). Generally a group should be cohesive and avoid the kinds of conflict that are destructive and disruptive, but healthy conflict and debate are important to decision making or problem solving. Passive group members may believe they are being cooperative and agreeable when conflict or a devil's advocate would be more valuable to decision making and problem solving. With open, critical, professional discussion and debate, the best ideas can be sorted out over time and the appropriate conclusions can be reached.

The group effort is no better than the composite efforts of the individuals in it. In working with a group, be critical and cooperative at the same time. A professional can be critical of ideas without being critical of the people offering those ideas. Cooperation demands that you be open-minded, that you listen as well as talk, and that you give and take opinions; it often requires that you allow your ideas to be criticized, that you criticize the ideas of others, and that you be willing to compromise on points of disagreement. The passive power of groupthink can obliterate the best possible ideas. Unanimity is a good quality. It should be strong once a decision is made, but it is a end, not an objective means.

Individual responsibility is important whether or not your group is working with an audience.

GROUP COMMUNICATION WITH AN AUDIENCE

An audience adds another dimension to group communication. Members of the audience are active participants even if they only listen. With the audience involved, group communication can take several forms. The **panel discussion,** carried out by a small group in front of an audience, is usually unrehearsed but need not be unorganized. The panel uses informal discussion with a leader to moderate the communication. It may appear to be spontaneous, but that appearance is usually the result of careful planning. A **symposium** is also made up of a small group, but participants act as individuals, with each making a prepared speech before the audience. Their speeches are on the same or closely related subjects. The symposium is more formal than the panel discussion, with little or no discussion among the participants. A symposium often turns into a panel discussion among the speakers or a forum with the audience. A **forum** involves all the attendants at a meeting, that is, total audience involvement, under the direction of a leader. The town meeting is a forum.

When an audience is involved, the planning for group communication is extremely important. Every group participant must clearly understand his or her role, how much time is allotted to the entire program, and how much each participant should use. It is equally important that members of the audience understand their role. Can they ask questions or make comments during the discussion, after the discussion, or not at all? Before any discussion of the topic, a good moderator or chair will explain to the audience and the participants the goals, procedural plans, and individual and audience roles. Determine your individual role as leader or as a member of the group, and take charge of your responsibilities. Let me illustrate these responsibilities relative to the panel discussion.

The Discussion Leader

If you are in charge of the panel as chair or moderator, **before the meeting** you will need to carry out several duties:

1. You may be responsible for choosing the panel members. Choose carefully. Participants should have expertise on the topic to be discussed, and they should be articulate and capable of presenting organized information. If a subject is controversial, they need to be able to maintain a professional attitude that does not allow for emotional outbursts or argumentative assaults on other group members. In other words, when you select participants, carefully evaluate their expertise, personalities, and communication skills. Although you may

rightfully have your own bias about a subject, balance the membership of the panel so that all sides of a question can be given equal consideration. Two conflicting personalities with two conflicting opinions can create a problem if both parties cannot conduct the discussion in an open, calm, and professional manner, especially if the discussion leader demonstrates a bias.

2. When you have chosen members of the discussion group or have accepted leadership of a group already chosen, you need to be sure you understand the purpose and the objectives to be put forth and accomplished. You may wish to consult with each participant for opinions to incorporate into the program. Study these opinions and your own carefully in view of the objectives and plan a strategy for conducting the panel discussion.

3. Once your members are chosen and your plans are clear to you, be sure that each participant understands the objectives, the procedural plans, and the role he or she is expected to play. Talk with each if possible, and send each a written agenda or outline of the program with names and roles of other participants and with suggestions for what you expect.

4. Consider the external influences on the communication effort. These include the audience, the room, and other physical conditions; the time involved and how portions of time should be allotted; and any other condition that could influence the success of your meeting. Sometimes bad weather, the noise next door, the time of day, or lighting can destroy an otherwise successful plan. Even the arrangement of the chairs in which participants sit will influence the meeting. A curve-shaped arrangement is often better than a straight line. Most people appear to be more comfortable sitting behind a table than not having a place for notes and arm rests. The room can be too large or too small, microphones may be needed, or your program may overlap in time with other sessions of interest to the audience. Your audience may be sleepy or belligerent. To whatever extent these things are under your control, control them.

At the meeting it is usually your responsibility as leader to take charge of the situation. Arrive a little early and take care of final details in the physical setting and greet the participants as they arrive. Your own confidence and relaxed attitude can be contagious. Carrying out the communication smoothly will also depend on how well you follow the organized plan. As leader, consider the following responsibilities:

1. Introduce the session to the group and to the audience if one is involved. You will introduce the panel members to each other, if they are not already acquainted, and to the audience. Then, you will outline the objectives for the discussion to both the panel and the audience. At this point be sure that members of the audience understand the role they are to play. Let them know whether you invite interruptions or prefer that questions and comments be saved until panelists have completed their initial comments. If you fear that the audience will be too vocal, you may wish to ask them to remain quiet throughout or to

submit any questions they have in writing (perhaps on cards that you provide). A smooth beginning can establish a rapport with the audience and contribute to a successful program. Despite your careful planning, when an audience is involved, be ready for surprises.

2. Your duties as leader during the session are to participate, moderate, arbitrate, evoke discussion, keep discussion on track, and summarize at appropriate junctures. Encourage all panelists to express their views; prompt the more timid with questions, and pull the conversation away from the aggressive talker. If an individual tries to dominate or if the subject gets off track, interrupt the discussion, summarize the points that have been discussed, and then redirect the group toward a question or a new point. If the discussion becomes too nebulous or general, tactfully call for details to support statements being made. In a relatively long session, it is good, even if the group is staying on track, to inject a summarizing statement more than once. This kind of control may be important in order to keep the program within a designated time as well as to be sure all issues in the original plan are discussed. Caution: In your role as moderator or leader don't monopolize the situation. Many leaders become too aggressive; it is your responsibility to guide and encourage the communication, not to subdue the group and do all the talking yourself.

3. You will also be responsible for concluding the session. Again, summarize the points discussed and any conclusions that have been reached. Often you will need to reiterate the objectives and show the extent to which they have been fulfilled. It will be important to include in your summary those unexpected points that arose during the session. In other words, you cannot completely plan your concluding remarks before the session begins; you must take notes throughout the discussion and quickly organize and present major points that were brought out. The job of concluding will be easier if you carried out your other responsibilities effectively.

Finally, in ending the session, you will include an expression of appreciation to the panelists and to the audience for their attention and participation. If further sessions are to be held or announcements made, you may be in charge of information about those. Always conclude with a professional and pleasant attitude as leader, even if the discussion itself has deviated from those qualities.

Responsibilities of Group Members

Leadership skills are essential for every panelist. Members must consider the audience, the individuals involved in the discussion, the objectives and plans for the meeting, the time involved and how much of it belongs to each of them, and the professional attitude essential for a successful meeting. As a member you may be expected to take the lead on certain issues, or you may be in charge of a segment of time. Often the designated leader or moderator may be put into an uncomfortable position by a question from the audience or a point being discussed about which he or she has previously developed a reputation. At such a point, a panel member can supply the needed support to keep the discussion on track by injecting an apt comment or temporarily and unobtrusively assuming psychological leadership.

All members, including the leader, must work as a group and as individuals. Each and all must be responsible (1) for defining, interpreting, and analyzing the subject and points that arise, (2) for keeping the discussion relevant and alive, (3) for evoking discussion from each other and, if applicable, from the audience, (4) for controlling unproductive conflicts and giving each group member equal opportunity to participate, (5) for giving each issue appropriate time and attention, and (6) for following plans set forth in the preparation for the meeting and helping to summarize the issues at various points along the way.

Planning for Discussion Groups

Organization for group communication includes the **subject matter,** the **time,** and the **people** involved. The following are possible structures on which you might base your plans for a group discussion. These examples apply where both a small group and an audience are involved, and they are certainly not inclusive of all possibilities.

The Forum. In a short statement, the chair may introduce the subject to be discussed and outline the goals to both the group (who should know already) and the audience. This leader then opens the discussion to both the panel and the audience. This technique turns into an open discussion among audience and panelists, or it becomes a question/answer session with the panelists serving simply as leaders answering questions and discussing issues along with the moderator. The method is valuable for public meetings in which experts are invited to serve as a panel of consultants. The danger in this structure is the difficulty in controlling a large group, especially if an issue is emotional or controversial and the audience includes some aggressive, opinionated personalities. All panel members need to help control the situation with their own attitudes and remarks and to move toward a prescribed goal in a professional manner. The leader should be a strong individual who does not allow the meeting to get out of control. Control in such situations comes more with finesse than with demands.

The Closed Panel. The chair may introduce the general subject in a short speech and then turn discussion over to the panel with no audience involvement except for listening and thinking. This technique is often used with television or radio broadcasts. The chair keeps the issues moving by asking questions and by directing and redirecting the discussion from panelist to panelist in an equitable manner. If a live audience is present, they may insist on being involved.

Panelists should be careful not to encourage audience involvement. Again, a strong leader must assert his or her position, outline the plan to the audience from the beginning, and divert any interruption that could keep the meeting from reaching a goal.

Symposium/Panel/Forum. The chair may introduce the general subject and the basic issues to be discussed. Then each group participant may also make a short speech in which he or she outlines one of the issues or one point of view. To this point, the format is that of a symposium, but it may develop into a panel discussion or a forum after the short speeches.

Decisions on which direction the meeting will take should, tentatively at least, be made before the meeting. Leadership roles may be divided among panelists by subject matter or points of view, and each of these leaders must assume responsibility for moving toward a common goal. Audience involvement may require that the leader serves as moderator between the panel and the audience. The symposium that develops into a panel and forum is valuable for controversial issues or for subjects that should be considered from several points of view. Success depends on a controlled, professional atmosphere.

Whether or not an audience is involved, prepare and organize materials for group communications as carefully as for individual presentations. A great deal of time can be wasted if discussion leaders are not prepared with an agenda or outline for conducting a meeting and if participants have not consulted that agenda and organized their thoughts and prepared supporting materials to express their views. Keep in mind several important points that can make group communication successful.

—**Set a specific goal but keep plans simple. With group interaction, you need to limit what you can accomplish in a reasonable time.**

—**Start on time, give each issue appropriate time, and end on time.**

—**Be sure everyone, including any audience, knows what format is being followed and what goal is pursued.**

—**Think as an individual; offer critical analysis and clear evaluation of information discussed; avoid groupthink.**

—**Work toward the prescribed goal, summarize along the way, and avoid digressions.**

—**Maintain a professional attitude and cooperate with the leader and other participants.**

—**Sustain equitable participation; don't talk too much or too little.**

—**To the extent that you can control such things, be sure the physical situation is comfortable for everyone and conducive to good communication.**

Recent technology has provided us with network discussions and conference calls that may or may not allow us face-to-face exchange with other members of the group. Group interviews may take place on television or radio with an audience receiving the broadcast but not in the studio, or a televised panel can appear before a live audience. You may find yourself interviewing for a job via conference call with a group on the other end of the line. Not having a direct view of the body language can be a handicap, but much can be expressed in the voice with tone, enthusiasm, and enunciation.

The group structures I have discussed clearly do not exhaust the creative possibilities, and your choice of format for communication may depend on the subject, how formal or informal you need to be, how much time you have, and at what point you wish to involve an audience. Whatever the situation, perhaps the most common group communication for the scientist is with a team of fellow researchers.

Whether it is a committee appointed to recommend a company policy or a panel of experts discussing an issue before an audience, the principles that take group communication to a desired goal are the same. Your participation should be active and ethical. You can work professionally with personalities in your group if you assume the proper role for your own personality and acknowledge the rights of others. Think independently and respect the independent thinking of others, even those who disagree with you. If you have trouble working with a group, study information on group communication such as that found in Beebe and Masterson (1994), Wilson and Hanna (1993), or Jensen and Chilberg (1991).

References

Beebe, S. A., and Masterson, J. T. (1994). *Communicating in Small Groups: Principles and Practices*, 4th ed. HarperCollins College Publishers, New York.

Janis, I. L. (1982). *Groupthink: Psychological Studies of Policy Decisions and Fiascoes*, 2nd ed. Houghton Mifflin, Boston.

Jensen, A. D., and Chilberg, J. C. (1991). *Small Group Communication*. Wadsworth, Belmont, CA.

Wilson, G. L., and Hanna, M. S. (1993). *Groups in Context: Leadership and Participation in Small Groups*, 3rd ed. McGraw-Hill, New York.

19

COMMUNICATING WITH
OTHER AUDIENCES

"We have a society exquisitely dependent on science
and technology, in which the average person understands hardly
anything about science and technology. This is the clearest
imaginable prescription for disaster—especially in a
purported democracy."

CARL SAGAN

In your scientific career, one of your most serious responsibilities is to communicate with those who know little about your area of expertise. The medical doctor and the science teacher face this responsibility every day. We sometimes complain about scientific illiteracy among the public, but not everyone can, or even should, have a degree in science. A specialist in the arts or humanities has no more time for studying science than you have for studying art and history. As Will Rogers told us, "Everybody is ignorant, only on different subjects." A little learning can be a dangerous thing, and often serious decisions are based on limited or inaccurate scientific information. It may be that those who are scientifically illiterate have not taken advantage of educational opportunities or have just enough learning to be confused. It also could be that scientists do not provide them with adequate opportunity to acquire unbiased knowledge. Keeping scientific information enveloped in an aura of mystery is not healthy. As any good scientist can attest, the unknown elicits curiosity, but it can also evoke emotions of anxiety and fear. As with other influences on our lives, science and technology involve risks. Help the nonscientist acquire enough information to satisfy some of the curiosity and to allay the fear.

In the first place, be proud of what you are and what you do. Progress in science is remarkable, and you want your audience to care about your science. To reach them, try to broaden your own mind and see values from their vantage point. Scientific and other human values often differ and sometimes even conflict. Science is not the only possible constructive truth in the world, and every scientific experiment is based on limited measures: "Any single set of factors can be interpreted in a variety of ways" (Gould, 1981). Science has produced both the good and the bad. Your belief in science can be firm, but someone else may believe as strongly that science is a wayward and destructive body of knowledge. Neither of you knows the full truth. Scientists are not amoral, and the products from science are not immune to other human values. Communication is the avenue through which you can reconcile differences so that those with differing opinions can work together for some specific good.

It is easy to ignore opportunities for communicating with nonscientists. You don't have time for auxiliary tasks that will probably provide little ammunition for your promotion. Besides, you have seen how "the public" can take a professional's words out of context and make legal lies out of accurate remarks by putting them in a new context. Certainly, scientific, and especially environmental, issues are not detached from politics and everyday life. Public policy on scientific issues often is shaped by the scientific illiterate. In a democratic society, anyone who knows about a subject should contribute to the quality and education of a self-governing citizenry. Theoretically, scientists should be the most objective, open-minded of teachers, but scientists have biases just as others do. Keep a high level of objectivity so that your communication with the nonscientist does not reveal your biases.

Whatever career path in science you take, you can find a way to convey some of your scientific knowledge to others. You can fulfill your responsibility for communicating with audiences other than your peers if you will think in terms of who makes up the audience, what avenues you have for reaching various audiences, how your subjects can interest them, and what techniques can best convey your messages to those people. If you enjoy writing and speaking as well as science, you might even look into a career in science writing, which can certainly involve oral science presentations.

AUDIENCES

Audience makes the difference in where, what, and how you communicate. How do you talk with or write for an audience that has no idea what deoxyribonucleic

"Excellent communication skills. Poor choice of words."

FIGURE 19-1

Adapt your choice of words to the audience. (Cartoon from Martha Campbell [1991]. *Phi Delta Kappan* **73**, 130. Used with author's permission.)

acid, trinitrotoluene, anthesis, or denitrification is? Above all, don't strut like a peacock simply because you know some words unfamiliar to your audience (Fig 19-1). You speak the same language they speak; you simply have acquired some alternative vocabulary words that make it more convenient to communicate with your peers in science. You should be capable of using a term like "flowering" or "bloom" rather than anthesis or of referring to "TNT, a highly explosive material" rather than trinitrotoluene. If you can't use such synonyms, perhaps **you** have a literacy problem.

I knew an advisor who told his graduate student just before a speech to "snow 'em with the terminology." He contended that, because he and the student were working in a new area of research, the student should be able to string together a vocabulary that even the scientists would have trouble following and, thereby, "leave them impressed." I felt at the time that that advisor should be beaten about the head and ears. In any communication effort, the objective is to make yourself understood, not to "snow 'em" with jargon and obscurity.

For a basic understanding of your audience, you need to consider such things as their ages and educational level. They may have never had an opportunity to attend college, and if they have doctoral degrees in the humanities, they may never have taken science beyond a general biology course. Most of them, however, are as intelligent as you. You need to think about their interests, their prejudices, their attitudes, and maybe their occupations. Find out as much as you can about their needs, their biases, their familiarity with your subject, and their reasons for wanting to hear or read what you have to say. Paint a picture in your mind of the typical person in your audience. That person could have high regard for your scientific education and respect everything you say; let that trust humble you a little. On the other hand, if the typical member of your audience has prejudices or a negative attitude about the values in science or your topic, he or she has reasons for such notions whether or not objectivity is at the root of those reasons. Try to discover the reasons and respect them as much as possible.

Your job will be easy if the audience is a homogeneous group such as science club students from grades six to eight or the chamber of commerce of a small town. More difficult is "the public" or even a college audience composed of graduate and undergraduate students from all areas of science. In your scientific paper written for your peers, you may say, or assume they know, that denitrification is the reduction of nitrates or nitrites to nitric or nitrous oxide by denitrifying bacteria under aerobic conditions with an available carbon source. You would be able to tell the college students the same thing, but freshmen might need some further explanation.

For the public, your objective will be different. They probably do not need a scientific definition; they need to associate the scientific idea with their own world. You may then explain that nitrogen fertilizers are essential to growing the food we eat, but too much nitrogen in the soil in the form of nitrates can get to the groundwater, which is the source of much of our drinking water. Too much nitrate in our water supply can be harmful to the health. To break down those nitrates, we need a soil that contains helpful microorganisms that convert the nitrates into less harmful gasses. With this information you are as close to the audience as the water they drink. Most intelligent adults will understand what you are saying whether or not they have sat through chemistry and microbiology classes.

For the kindergartners, you might need to explain even further; the microorganisms may become tiny bugs that we can't even see. Whatever adjustment you have to make in your terminology, there is no reason to suppose that you cannot communicate with a group of people simply because you are more educated in one area than they are. Someone in your family or among your acquaintances fits into the audience you must address. Picture that person and think about how you would tell him or her what you have to say.

AVENUES

Identifying the audience and empathizing with them are keys to success with any communication, but how do you acquire such an audience of nonscientists?

Opportunities will arise, but you can also open doors yourself. You could join civic groups that would be pleased to have you provide scientific expertise on public problems such as safe water supplies or waste disposal. You may belong to an interest group such as amateur bird watchers who would like you to tell them more about birds, ecology, or other areas that you know better than they do.

Your own company or agency could provide avenues for communication. Your position can require that you write for or speak with lay audiences. A major responsibility of scientists in the Agricultural Research Service and the Cooperative Extension Service is to transfer information about what is going on in scientific research to the agricultural producers and industries. Your efforts may be to inform nonscientists within as well as outside your company. You can make speeches or produce copy for fact sheets, leaflets, technical bulletins, newsletters, policy statements, employee or public pamphlets or brochures, exhibits or posters, and other creative forms. Your employer may encourage you to visit civic clubs or schools or write for the popular press.

The media come in numerous other forms. Often scientists are invited to participate in interviews or other programs on radio or television. Live interviews are a bit scary at first, but good preparation can alleviate the fear. Suggestions by Gastel (1983) can help. Rather than the live interview, you might be more comfortable with other avenues for communication. You might produce your own videotape that can be edited for a given audience such as high school students, first-graders, or the public. You can be invited to give a slide presentation, give a demonstration, or make a speech to a group of children or a civic club. Electronic communication can now reach many audiences. Children of the future will use a computer network to acquire information that we used to obtain from the encyclopedia. You could contribute to a store of information on that network.

The written media contain even more variety than the spoken. Write for a newspaper or popular magazine; you can submit a feature article or maybe contribute a regular column. Some scientists even pursue careers as "science writers" for the popular press. But you can maintain your research, teaching, or service position as a scientist and still use the popular press and other media for communication with nonscientists. Both in speaking and writing, you will find great opportunities for alleviating some of the scientific illiteracy that prevails.

SUBJECT

The subject you select or are asked to write or speak about will depend on the audience. As a scientist, it is in your own best interest to provide information that will help nonscientists understand, trust, and appreciate science. They certainly have a right to know how seedless watermelons have been developed, what chemicals are in their food, or why a hormone supplement can make them feel better and even change their attitudes. They also have a right to know whether genetically altered species might destroy Mother Nature's own creations, why we should bother to protect a blind lizard or a night bird, how to effectively dispose of waste that humans and animals create, whether nuclear energy is as safe as electricity, what side effects drugs may produce, and hundreds of other questions that affect their daily lives, their jobs, and their government. Help them to know the real risks and any safety precautions needed as well as the benefits.

Try to select a subject based on the interests of the audience. It should be timely and important. Clearly identify your objective or purpose in communicating your subject. You may tell a group what they can do to restore their land to productivity after a lengthy flood or help them understand how to guard against communicable diseases. You could encourage them to read and follow pesticide labels before scattering herbicide in the yard or insecticide in the house. Just like you, they are interested in safety, current happenings, unusual events or discoveries, and new developments in science and technology. They are interested in nature and in using and protecting the environment. They care about what science does to help or hurt the economy, their employment, and their quality of life.

How you treat the subject can be as important as the subject itself. Be cautious with disclosures about new discoveries. Better to tell the audience you don't know than to lead them through your own enthusiasm to making faulty presumptions based on what is still experimental. Often when you do not wish to commit yourself to a possible outcome of a scientific discovery, you can refer the audience to original sources of information and let them draw their own conclusions just as you do. Remain positive and congenial if they question your credibility. Don't assume that the audience is unintelligent or has a bad attitude. Principles for good communication require that you respect your audience and take criticism, perhaps from people that know far less than you do about your subjects. React to that criticism in a calm, rational manner. It can help to have someone review what you will say or write for a given audience and then make revisions before you deliver the information.

TECHNIQUES

With a clear purpose and your audience in mind, your task is to relate the unfamiliar to the familiar. You have to start with what the audience knows already whether they be high school science students or preschoolers at your daughter's day care center. You will need to use more extensive illustration than you would with your peers in science. Slides, demonstrations, and other displays can help them understand your points more quickly than words alone will. Don't hesitate to use scientific data for support of what you say or write. Just be sure to present those data in a form that your audience will understand and that is credible and not fragmentary. Use definitions and comparisons and visual imagery. Compare ideas in the science with something familiar to the audience. For example, if we could see a DNA molecule, it might look like a couple of chains looped around each other. Third graders could visualize that image, especially if you present a picture.

Writing or speaking about science should follow the principles of simplicity as laid down by such authors as Zinsser (1998). Be sure that what you say or write is accurate and as complete as possible. Be conversational and direct, and repeat your main points for emphasis. Use first and second person and the active voice. And watch your tone as well as your vocabulary; watch out for any note of prejudice or condescension in your diction. It is easy to inadvertently say something that can offend some individual or group in your audience. Avoid complexity and incomplete explanations. Be sure you carry the audience forward with transitions that they understand; these transitions sometimes require more lengthy explanation of what connects two points than would be required with a scientific audience. Narrow your subject to one that can be explained in terminology the audience will understand in the speaking time or writing space you have available.

In addition to conveying the unfamiliar in familiar terms with accuracy, completeness, and simplicity, consider the interests of the audience. What they care about knowing is as important as what you think they should know. They may not care what a DNA molecule looks like or how an *Agrobacterium tumefaciens* can serve as a vector to transfer DNA from one species to another, but they want to know how you can use knowledge of DNA to change characteristics in a tomato or provide evidence in a rape trial. Join them in their world; don't expect them to come to yours immediately. As a good communicator, you can guide them to at least one view of your world if you lead them from their own interests.

Several sources in addition to Zinsser (1998) can be helpful for exploring the principles of good communication. Gastel (1983) provides good information on media and techniques. But you also may need to return to basic instruction in a text such as that of Burnett (1994) for writing or Smith (1984) for speaking.

Read some of the popular magazines or newsletters in your discipline; you can find many of them on the library shelves next to your professional journals. Read and listen to Carl Sagan, Stephen J. Gould, or other well-known science communicators.

SCIENCE WRITING AND PRESENTATION

Much more is said about science writing than about science speaking, but both are important. As Day (1995) suggests, a distinction needs to be made between science writing and scientific writing. He points out that this distinction lies in who the authors and audiences are. Scientific writing is done by scientists for an audience of scientists. Science writing is done by scientists or journalists for an audience of nonscientists. I would add the important differences in the treatment of subjects and techniques of delivery to Day's distinction.

Certainly science writing and presentations on science can constitute a career in itself. For any of you who may be interested, look into the book edited by Lutz and Storms (1998), especially Chapter 7 by Ricki Lewis. You may want to investigate the National Association of Science Writers (NASW); maybe start with an Internet search at http://www.nasw.org. Blum and Knudson (1997) have edited *A Field Guide for Science Writers*, the official guide of the NASW. It contains commentary by well-known science writers about the media and audiences they write for and the subjects they write about. Gastel (1983) provides good commentary on writing science for the public. Also look at examples of science writing. Gannon (1991) provides examples along with commentary by the author, and Zinsser's *Writing to Learn* (1988) contains good examples of science writing. In *Headline News/Science Views*, Jarmul (1993) shows a wide range of science writing on current topics that make the news. Of course, anyone who wants to be a science writer should read various books by such well-known science authors as Stephen J. Gould and Carl Sagan; they can teach you much by example.

You can have an interesting career in science writing, but let me reiterate that this science communication extends to oral and visual media as well. For many years, Jacques Cousteau taught us a great deal about marine biology on television, and many shows still offer entertaining and educational science. Speaking engagements, lecture tours, use of the Internet, broadcast journalism, and other avenues can allow you to present oral and visual information to audiences that may be more receptive to listening and seeing than to reading. Whatever form you use, what is important is that you are informing and educating nonscientists about science. Find out where the people in your neighborhood are getting information about science. Then make a commitment to communicate science outside the scientific community.

You don't have to be a Jacques Cousteau or Carl Sagan or Stephen J. Gould. You cannot provide all the answers to all the public; but as Richard S. Nicholson says, "Rather than regard such activities as irrelevant distractions, we must realize that informing the public is a vital aspect of our jobs." If each practicing scientist would contribute a bit of his or her knowledge to nonscientists, scientific illiteracy would not be such as problem. In fact, it could well be that we would all take better care of that world of physical matter that is the subject and breeding ground for science itself.

References

Blum, D., and Knudson, M., eds. (1997). *A Field Guide for Science Writers*. Oxford University Press, New York.

Burnett, R. E. (1994). *Technical Communication*, 3rd ed. Wadsworth, Belmont, CA.

Day, R. A. (1995). *Scientific English: A Guide for Scientists and other Professionals*, 2nd ed. Oryx Press, Phoenix, AZ.

Gannon, R., ed. (1991). *Best Science Writing*. Oryx Press, Phoenix, AZ.

Gastel, B. (1983). *Presenting Science to the Public*. ISI Press, Philadelphia, PA.

Gould, S. J. (1981). *The Mismeasure of Man*. W. W. Norton & Co., New York.

Jarmul, D., ed. (1993). *Headline News/Science Views II*. National Academy Press, Washington, DC.

Lutz, J. A., and Storms, C. G., eds. (1998). *The Practice of Technical and Scientific Communication: Writing in Professional Context*. Ablex, Stamford, CT.

Smith, T. C. (1984). *Making Successful Presentations: A Self-Teaching Guide*, 2nd ed. John Wiley & Sons, New York.

Zinsser, W. (1998). *On Writing Well: The Classic Guide to Writing Non-Fiction*, 6th ed. HarperCollins, New York.

Zinsser, W. (1988). *Writing to Learn*. Harper & Row, New York.

20

TO THE INTERNATIONAL
STUDENT

"We can work together, men and women, to develop the world."

MARIA MASHINGO

Although I have tried to avoid idioms that are too colloquial in this book, for authors or speakers whose first language is not American English, I am sure that some of the things I have said earlier are perplexing. The lack of clear communication between us is the result of my language adhering closely to my personality and my culture and your understanding being based upon your culture and personality. Your English and my English can both be high quality and still be influenced enough by our cultural backgrounds that it is sometimes confusing to understand each other. In this final chapter, I would like to discuss some of those cultural issues that impact our writing, speaking, body language, and even our opinions on what should be communicated and how to communicate effectively.

First, I think we have to look at some cultural differences that influence our beliefs, our behavior, and our life styles. All these issues affect the way we write and talk. Then, let us look at the expected conventions for speaking and writing in American English, including such issues as body language and plagiarism that differ in many cultures.

The opinions I express here may or may not be entirely fitting for your situation. You may have studied English since preschool or you may have just begun to learn English as you have gotten ready for graduate school in the

225

United States. You may have traveled a great deal in the United States and already have a clear concept of our culture. In addition, I am no expert in cultural differences. What I express here is based upon my own observations, the literature I have read, and the discussions I have had with international students at my university. Here, I will refer to the persons from other nations as internationals and those from the United States as Americans. Both these terms are too general for the diversity of peoples included, but for a single chapter on this complex subject, generalizations are necessary.

BECOMING ADJUSTED TO U.S. CULTURES

The world is too big and complex for any individual to see it from all directions and with total comprehension of its physical, metaphysical, and even theoretical contents. From the physical or empirical world, as Hall (1983) suggests, we create much of our own environment by building beliefs and physical and social structures to accommodate a way of life. In doing so, we develop societies of people whose speaking and writing are influenced by their natural environment and the structures, values, beliefs, customs, and activities they put into it. All these things contribute to a semantic environment in which a language evolves. At points at which cultures are similar, we understand each other. Where they differ, we are often frustrated with misunderstanding.

Fortunately, for scientific communication, a common environment of natural phenomena and common techniques for defining empirical behaviors in that environment make it relatively easy for persons from different cultures to communicate about science. However, for students coming to the United States from other cultures, it is impossible to separate communication that is related to your entire experience in this country from that which focuses on scientific papers and presentations. Consequently, an understanding of the environment in the U.S. college community and some of the language differences as influenced by our cultures may contribute to your success with scientific papers and presentations for American audiences.

What are Americans like? They have originated from many cultures and are too diverse to describe. When you come to the United States for the first time, keep your mind open about what you think of the behavior of people here and be flexible in your opinions. Most students and professors in the college communities where you will be living are open-minded and relatively unbiased. As elsewhere in the world, we have our individual personalities; one person may smile easily, another may almost always have a worried look or a frown on the face. Your culture is as unfamiliar to most of us as ours is to you. This lack of familiarity may create a distance between us at first. One student told me about going to classes and to meetings for the first few times in the United States and feeling that others seemed to avoid sitting beside her or talking with her. Although unlikely, that behavior could have resulted from prejudice, but it may well be that the U.S. students saw familiar faces in the room that they wanted to join. Very soon this student made her own efforts to talk with others, found that most did not avoid her, and felt comfortable in dialogue with Americans.

Keep in mind also that the U.S. student is probably more ignorant of your country than you are of ours. Unfortunately, in this country we do not teach most of our students enough about other cultures, other societies, and other countries. Most of them have never lived in or even traveled through another country. Their opinions may be generalizations formed from movies or news reports that do not depict your country as it really is. Be patient with us. When you have an opportunity, tell us about the people, the customs, the cultural differences in your country so that we can learn from you and adapt to you as you are adapting to us. At the same time, any preconceived ideas that you have about people in the United States may not be entirely true. Be open-minded to whatever you encounter. We are a mix of beliefs, cultures, and social and economic conditions. Movies that you may have seen, news reports you have heard, or music you have listened to will not tell you what the United States is like. We are not all lovers of rock or country music. We do not all have good cars and spacious houses, and not everyone has a good education or a good job. And we do not all speak English with the same accents and fluency.

You may fear that you will have trouble understanding our English or making us understand yours. The fact is that, across this country, we sometimes have trouble understanding each other's English. The English of the African American from Mississippi and that of the Irish New Yorker may well be as different from each other as yours is from either of theirs. Don't worry; you will adjust to the accents in the area of the United States where you live. Especially at first, we may ask you to repeat or rephrase something you say, and don't hesitate to ask us to repeat. But as our ears become accustomed to your sounds and your ears to our sounds, we will understand each other quite well.

Communication is closely associated with personality of the individual, but that personality is often steeped in unique cultural and environmental backgrounds that we may not even be conscious of in ourselves. You may find that it is frustrating to adapt to the new physical environment where geography, climate, and even food and clothes are different. Students from some countries are sometimes amazed at how large the United States is. They may know other students or relatives from their countries who live in the United States, but they find that visiting someone in Oregon from a graduate school in Florida may involve traveling over almost as much distance as traveling back to Sri Lanka or Mali. Public transportation systems are not always convenient or may even be nonexistent in smaller cities or towns. As soon as you arrive or before you arrive in this country, try to study the particular place where you will be located. In addition to transportation, the foreign student in the United States may be frustrated by the living conditions, the climate, and access to social and religious activities that they want to be involved with. Depending on the part of the United States you are in, the climate may be hotter or colder than that in your country, and it may be difficult to purchase the kind of food you are accustomed to eating.

You can find a great deal of information through the Internet. The state or the city in which the university is located and the school itself will probably have Web sites. Talk with other international students who have lived here for a while, especially the ones from your own culture. Temporarily at least, you must adjust to the culture and the language in order to be successful. That is not to say that you have to *adopt* the culture, just *adapt* to it for the time you are here. We call the United States a melting pot of ethnic populations, and it is made up of numerous cultural influences from various places in the world. We don't always agree with or understand each other very well. But with this diversity, most of us try to tolerate and understand differences and do not reject people because of cultural differences or beliefs. Most of us welcome new ideas. Although it is not characteristic in a college community, you will probably meet individuals or groups who have ethnic, religious, racial, or socioeconomic prejudices. Such people may exist in your country too. Please don't assume that these people are representative of the basic U.S. principles and beliefs. And don't apply the following generalizations entirely to yourself or to other individuals that you meet. Every individual is different.

GENERAL CULTURAL DIFFERENCES

Attitudes Toward the Self

In some cultures, the individual sees himself or herself as a member of a group, a family, a neighborhood, or another social entity and is more interested in the preservation, well-being, and harmony in that group than in his or her own self interests. In the United States, the individual is often the focus, and allegiance to a group typically does not take precedence over self-interests. This attitude can be interpreted as selfish and arrogant or as cold and too assertive. Most U.S. citizens see it as independence, freedom of choice, and respect for the worth of the individual. The American may see the international as too pliable and not individually assertive enough. It should not be too difficult for each of us to maintain our own perspective and to respect that of the other.

Direct and Indirect Communication

Another difference that Edward Hall and other linguists and anthropologists observe is that some cultures are "high context" and some are "low context." As

I understand this concept, in low-context cultures such as those in the United States, Denmark, or Germany, the communicator is very direct and does not depend so much on the obvious context of the environment or a situation to express an idea. The direct words and body language express what he or she observes, even if the listener is observing the same thing. In the high-context cultures often found in the Far East or Latin America, the communicator does not feel the need to express what is already obvious in the context of the situation but provides a key expression or hint that allows the listener to make an interpretation of meaning beyond the obvious. One student from Latin America described the attitude of the Danish as rude in that they, in effect, told her that if she didn't like their country, she could just leave. She would never have voiced such a suggestion to anyone; that option would be obvious in the context of the situation, and it would be offensive to voice it.

These attitudes are often described as direct and indirect communication and affect speaking and writing as well as behavior of the individual. Behavior and attitudes of the American may be seen as rude or uncaring by some internationals but are considered reasonable by persons in the United States and other low-context countries. Those of the international may be interpreted as weak or insecure by some, but others may interpret the behavior and attitudes as simply adhering to a principle of harmony. Along with direct communication, the American is often uncomfortable with silence and will fill a silence with words even if they are repetitious or ineffective. The indirect version of communication welcomes silences as time to think and interpret what is being said and not said. Although it is not likely to happen, perhaps we could all modify our behaviors by mixing the direct and indirect appropriately.

Power Distance

Complementing allegiance to a group or an individual and to direct and indirect attitudes is the concept of power distance. In groups, there are various positions that individuals hold such as the parent or eldest son in a family, the boss or coworker on the job, the teacher or student in the classroom. These positions entail social or professional conventions that individuals adhere to. Behavior and communication depend on how respect is expressed for the various positions in the group. For example, in some cultures children are expected to remain silent in the presence of most adults, or students are expected not to express opinions contrary to the opinion of the teacher. These practices are conducive to group harmony. In the United States, we generally encourage children or students to voice their own ideas both at home and in the classroom. At school this concept lends itself to the use of different teaching techniques, and the international student accustomed to being silent in class may find it difficult to speak out when expected to join a class discussion. The international may see students in the United States as rude and disrespectful, and the American may see the international student as being too timid and reticent. Even professors may misinterpret silence as a lack of knowledge when, in fact, it may indicate clear understanding.

Time

The cultural differences in our concepts of time and its importance are interesting. Our ideas of the priority time has in our lives, the importance or lack of importance of punctuality, our management of time or schedules — all these issues are often different for different cultures. Treatment of time may stem from whether we give more priority to strict scheduling of events or to choosing to act when the time is most convenient or appropriate. In the United States and in other low-context countries, adhering to schedules is very important. Being late or early or taking more time than is scheduled for an event is generally offensive. Typically, Americans are rather flexible relative to most ideas, but you may find them intolerant of a lack of punctuality. Most have great respect for schedules, and professors may set deadlines for assignments and not understand a cultural lack of punctuality in meeting those deadlines or in your coming early or late to a class or an appointment. An excellent paper turned in late may be graded down simply because of lack of punctuality.

Which is best — the greater or lesser power distance, the direct or the indirect communication, the group or the individual focus, or strict or loose adherence to time schedules? Our cultures dictate the answers for us as individuals, and it is certainly not the purpose of this chapter to debate the pros and cons of the different behaviors and communication styles. But understanding that these differences exist may explain why and how we behave or communicate differently. Our purpose here is to determine how these cultural differences affect scientific writing and speaking and to make all of us more comfortable with any audience. Your goal as an international has to be to make whatever adjustments are needed to be successful in your work here and in your speaking and writing American English.

SCIENTIFIC WRITING IN AMERICAN ENGLISH

If you are an international student from a culture where indirect language is preferred, you may find that your reading audience, whether professors, other students, or journal readers, may think your literary style is too diffuse or rambling. Students tell me, and Fox (1994) says her students tell her, that professors often instruct international students to "get to the point" and be more concise and direct with support for a point. Writing by both American and international students is often redundant; is filled with long, wordy, complex sentences; and lacks clarity and organization. A direct, simple approach can help to avert these deterrents to understanding. I certainly will not argue about whether a forceful, direct approach is superior to other forms of communication, but that approach does lend itself to simplicity and is the form preferred for scientific audiences in the United States.

If you are from a culture that does not use the direct technique, the adjustment in organization and development of a topic may require that you add a pattern of thinking to your education. My contention is that the truly educated individual is the person who can and does communicate well with diverse audiences. Patterns in thinking constitute the base for our patterns of expression. Maintain your own patterns in the process; they will be important when you communicate with the audiences in your culture. But while you are in the United States, adapt to the organizational style understood by this audience. Developing new ways of thinking and communicating is difficult. Montgomery (2003) has suggestions for adapting by using models of English papers for developing skills in reading, speaking, and writing. I would add listening to those skills. For writing a paper, it may be helpful to read Chapter 3 of this book and study the example in Appendix 3, but let me expound upon some of these ideas more specifically here.

Because some of the difference in direct and indirect cultures and communication is in the organization of thoughts, probably the best way to learn a new communication technique is to begin with an outline or a listing of points you need to present and develop in your paper. Know that, for scientific writing in American English, the main point comes very near the first, and then you can give information about and support for that point as you develop the paper. We think of this technique as logical and reasonable, but it may be no more valid than a method that considers various background points of support and leads finally to a main point or conclusion. Again, our purpose is not to determine superiority of one method over another but to make it easier for you to communicate successfully with the American audience. The following is a discussion of the pattern for a scientific paper in American English. The suggested content for each part is based upon a typical journal manuscript. If you are writing a book or a thesis, the order may be different; if you are writing an abstract or a short report, each part may be shorter. But in any kind of paper, the American audience expects that you will make your point clear from the beginning and then develop information about that point concisely with the supporting details. In addition to the following information, look also at Appendix 3 for an example of an overly simple outline and paper that follow this pattern.

Introduction

Your paper may require an abstract or summary before the introduction, but the paper itself will begin with this section. The introduction for a journal manuscript will typically do three things: (1) almost immediately call attention to and define or clarify your specific topic for the audience; (2) provide brief, precise background necessary for understanding the topic and justifying why you are working with it; and (3) clearly define your main focus or objective relative to the subject. In an abstract, these things would be done briefly with a sentence of rationale and definition that takes care of the first two things and then a specific statement of objectives. In the journal manuscript, in about two double-spaced pages you may define and justify the subject, provide background information, make reference to the literature showing what other researchers have found, and then suggest a basic hypothesis and specific objectives upon which your research is based. If you find direct communication is foreign to your way of thinking, this introduction is the part where you may ramble or not get to the point unless you curb your desire to discuss too much background or too many ways of approaching your objectives.

Body of Paper

The main body of most scientific reports is composed of the materials and methods used in the research, the results acquired, and a brief discussion of the meaning of those results relative to what your initial purpose or hypothesis was and relative to what others have found about the topic previously. The materials and methods or experimental procedures section tells other researchers anything they would need to know to perform the research just as you have. An easy way to organize this section is to arrange it by objectives and describe the techniques you used to test each one.

The results should be a clear presentation of representative data from your study, probably much of it in the form of tables and figures. However, some results can and should be stated more simply in just the text, and the text should directly point to the significant findings depicted in the tables and figures. The discussion may be interwoven with results, or it may be a separate section. As with the introduction, take care not to ramble away from the main point in the discussion; just associate your interpretation of the meaning and significance of your results with scientific principles and reference to what others have found.

Conclusions

A conclusion to your paper should briefly reiterate specific points about what your objective sought to accomplish as compared with what was actually accomplished without detailed repetition of results and discussion. It may then show how the results from your work can be used in future research or in practical application. Finally, at the end of your manuscript, you will attach a list of references and any acknowledgment that is needed.

This organization and the content development of a research paper for journal publication are described in more detail in Day (1998) and are illustrated in Appendix 3. Similar direct patterns of organization and development can be designed for a short essay (Appendix 2) or for the thesis and other research reports. The proposal will follow this pattern except for omitting the results and discussion or including them only as preliminary findings. See the chapters in this book on theses and proposals and study the proposal in Appendix 5. We use a similar direct approach for writing reviews or brief communications such as letters and notes, but of course, the organization and content will depend upon the form. For all kinds of writing, reading some of the experiences of international students who have worked with Fox (1994) may also be helpful if you are frustrated with adjusting your ways of thinking and writing for an American audience.

Plagiarism

In both speaking and writing, an important issue in the United States is the concept of plagiarism. I have been told that in some countries it is a gesture of admiration and respect to another author to use his or her concepts and words in your own speech or paper and that you may often do so without noting specifically where your words and those of the original author are the same. In the United States, we consider the tangible expressions of an author "intellectual property" and must ask permission for or acknowledge use of it. This intellectual property may exist as a concept described with words or in other media such as data in tables and figures, choreographic or musical scores, or other tangible forms of expression. Use of intellectual property by others can be considered theft and is unethical and even illegal if it is not clearly documented or permission is not obtained.

Of course all of us use information from others, and if it is a few short lines, it can be paraphrased with different words or put in quote marks if the same words are used, and then the source is clearly cited. If longer passages of information are used, one should determine who owns the copyright on the words and obtain copyright permission before using them. Sometimes permission is freely granted, and sometimes an author will charge a fee for using his or her words. For example, for Appendix 7 here, Dr. Vander Stoep gave me her permission with no charge or restriction, but for Appendix 12, I paid the publisher who owned the copyright for Imhof's work. In both instances, I have retained the letters I wrote and the written permission in their replies as evidence that I have not stolen intellectual property (see Chapter 12).

If you are accustomed to the practice of using other sources freely without clear documentation or permission, you need to be very careful that you adjust to the system of documentation in the United States. We also find unacceptable the use of too much information from a single source or use of your own words that you have written for an earlier paper without referencing that paper. This may seem a simple matter to you, but it is very serious with work in the United States, and professors may fail your work if they detect plagiarism. For your own sake, try hard to understand the concept and adhere to it while you are working here. I have a little information on plagiarism in Chapter 12. You can also explore the Internet for definitions and examples of plagiarism on such sites as http://www.zoology.ubc.ca/bpg/plagiarism.htm#sel (Gordon *et al.*, 2003).

ORAL PRESENTATIONS

Even more than in writing, cultural background and personality influence the spoken language. If your knowledge of English is basically good — and for most international students I find that it is — then just give yourself a few weeks in this country and your ears will be attuned to what you hear and you will speak the language quite well. Some of my students worry about their accents. I tell them that pronunciation is important, but they should keep their accents and be proud of them. Accent is a part of pronunciation, but the sounds include tone, inflection, and variations in the rhythm and pitch of the voice, as well as your distinctive manner of expression, all of which depend on the phonetic qualities of your first language. If you travel in the United States, you will find that we also have various accents in different areas of the country. The Bostonian may have problems understanding the person from Georgia or Louisiana. Your accent is likely no more distinct than those. If you are careful with basic pronunciation and speak distinctly and slowly enough, you will likely be understood.

Articulate the basic English sounds as clearly as you can. Some of those sounds may not be present in your language, or one sound is almost indistinguishable from another for you. For example, the sounds of *s* and *th* are apparently similar for the Chinese. One young man from Saudi Arabia told me that the sounds of *b* and *p* are indistinguishable in his language. At a parking lot in the United States, he asked a policeman if he could "bark here," and the policeman smiled and said, "Of course, you can bark anywhere you like." Unfortunately, when the young man returned to his parked car, he had a ticket. He quickly learned to make a distinction in the sounds of *b* and *p*. Listening, reading aloud and recording and listening to yourself, along with thinking and talking in English will develop your ability to speak English well.

These listening and speaking exercises will also help you develop vocabulary and learn idiomatic expressions and sentence structure. The order of words, the use of articles and plurals or other word forms, and the construction of sentences may differ in English and your first language. Some languages do not associate gender with pronouns, and to distinguish *he* from *she* for men and women may be difficult in English. In other languages this difficulty increases in the use of gender with nouns naming inanimate objects. In Spanish a house is feminine, and a car is masculine. Your language may not use what we call articles (*a, an, the*) in front of nouns. English is inconsistent in this use, but some words sound strange to us without the article; which article you use or emphasize can alter the meaning of an expression. You learn such things about English as you experience it, but becoming fluent is a matter of practice. As with any sport or other physical skill, continual exercise and practice is the only way to become proficient. Practice the right pronunciation and the right order of words and learn to relax, and you will master the language. If you don't practice or if you repeatedly practice the wrong action, weaknesses will become difficult to overcome. Above all, do not resort to spending most of your time with others from your own culture and speaking your own language with them. Even if you speak English with them, you are not giving your ears a fair measure of American English. Associate with your American colleagues. Some international students even encourage their American friends to point out their weaknesses in pronunciation and vocabulary. In turn, they sometimes offer to teach the American some of their own language. Such an exchange is mutually beneficial and semantically healthy.

As in writing, some Americans are often very direct with their spoken words and make a main point clear to begin with and then discuss issues involved. Depending upon your culture, you may put background and discussion first and then end on your main point as you may do with writing. In making speeches or presentations, it is important that you adjust your thinking enough to adapt to the American pattern so that the audience will follow your talk better. Refer to Chapters 13 through 18 for other information on oral communication, but let me use this opportunity to reiterate some of the important issues.

Speed, pace, and volume are all important to good speaking and audience comprehension. I consider speed and pace as different but related qualities. Obviously, if you talk very rapidly in any language, it will be more taxing to the listener to keep up with what you are saying. Some of you may have developed rapid speech because of the influence of your own language. Sounds in some languages are formed in the front part of the mouth and can be emitted far more quickly than those sounds that originate deep in the throat. For example, many sounds in Spanish originate in the front of the mouth and are quickly emitted; many in English come from Germanic languages, in which the guttural sound originates deep in the throat and emerges more slowly. I find that many of the students whose first language is Spanish speak English rapidly. However, if this speed is modified with the pacing of words and phrases, language is easier to comprehend. Pace involves pauses after phrases and clauses. A phrase or sentence may be spoken rather rapidly and still be understood clearly if it is followed by a pause. These pauses allow the listener to catch up and keep up with sounds. They are very important for speakers in any language. Both American and international speakers should be conscious of speaking at an appropriately slow speed and pacing their delivery well to give the audience time to think.

Volume can be a problem for speakers of any language. Seldom is the problem one of talking too loudly, although a few speakers do so and appear too aggressive. For many international students, the reverse is true. Power distance or harmony issues relative to age, prestige, position, or gender may have caused you to develop a soft low tone and volume to your voice. This tone is usually pleasant, but the volume must be great enough that it can be heard. In a large room it is essential that you project your voice so that everyone in the audience can hear you. If you have a low voice, use a microphone if it is available. If not, you must make every effort to increase the projection of your voice to accommodate the situation. You may as well not say anything if what you say is not heard. All these remarks are true for the American English speaker as well as international speakers. Nervousness in front of the audience, a feeling of insecurity or inadequacy with the language, or timidity and cultural issues can make you talk too fast, fail to pause, or speak too softly. Be conscious of any of these problems you may have and remedy them.

CASUAL CONVERSATION

International students can often adapt rather easily to speaking situations such as making a presentation before an audience. More frustrating, especially in your first days or weeks in the United States, may be feeling comfortable with casual conversation, especially with such things as greetings or use of names and titles. We have many ways of greeting each other from "hi" or "hello" to "good morning" or "good afternoon." Sometimes the greeting is in the form of a question as in "how are you?" or, more casually, "what's happening?," "what's up," or "hey, what's new?" Don't attempt to answer such questions specifically; simply acknowledge the greeting with a general comment such as "I'm fine" or "not much new." Any of these greetings can be appropriate at the right time; listen for clues to when Americans use one or the other greeting, and until you become comfortable with them, you will be safe to simply say "hello" or "good morning (or afternoon)." A smile is almost always appropriate with a greeting.

At first, you may also feel uncomfortable with the way you address others, especially those who have titles. And names are treated differently in different cultures. Some international students are afraid they may not address their teachers appropriately. Again, in the United States, we have no single right way to address them. Let's look at an example of someone named Mary Jones who is a professor with a doctoral degree. Some professors prefer that their students, especially graduate students, call them by their first name; for example, Mary. Many prefer to be called Dr. plus their family name, as in Dr. Jones. At some schools it is still fashionable to refer to her as Professor Jones. My suggestion is that you call your professors, Dr. plus the family name (Dr. Jones) until you determine his or her preference.

Slang and American English idioms and metaphors can be confusing. You will also get more of these than you need by listening and talking with Americans. However, it will be just as well if you avoid much of the slang and meaningless expressions, especially in speech making. Most of the time you should avoid such unnecessary insertions as "OK," "Umm," "like," "stuff like that," "you know," "well," and "yeah." Crude, off-color, offensive slang should especially be avoided. Just because some Americans use those terms does not mean they are effective. Many of us are careless with our use of language. Do not use an expression until you know what it means to your audience.

BODY LANGUAGE

In speaking, perhaps body language and what is unspoken are more important than the words spoken. In fact Hall and Hall (1990) say that 90% of our communication with each other occurs beyond the words we speak. The expressions on our faces, the gestures we make, the way we respond to silences, the clothes or jewelry we wear, or our hair style—every element in our interpersonal contact says something to the audience. You will learn most body language by observation the same as you learn other meanings in language. Some body language is fairly universal, but even a smile or frown or a nod of the head may be interpreted in different ways. A smile, for instance, certainly has a similar meaning worldwide, but one student from Serbia told me that Americans smile too much. She resented a speaker who "smiled too much" as he presented serious scientific data. Her personality as well as her cultural background probably influenced that opinion. Lightly shaking the head side to side in the United States means "no" or that you disagree with the speaker, but Indian students have told me that it indicates agreement in their country.

Some body gestures are more distinct for different cultures than the smile or a shake of the head. Generally Americans and internationals accept differences in most of these behaviors and can easily coordinate meaning without misunderstanding. For instance, if your culture prefers that your clothes be different from what we see as the usual American attire, most of us find it interesting but not objectionable, and we will see beyond that difference to interpret your communication appropriately. The same is true of a gesture such as a bow that may be common in your greeting of a professor. English and American people used to bow, especially relative to power distance, but as we have shortened the power distance in American society over the years, we have essentially lost the art of bowing. Still, it is generally pleasant to have someone from a different culture bow when he or she is greeting us. When he first arrived here, a young man from Korea bowed his body from the waist when he spoke to me in the hallway; within a few months he had modified that by simply giving a bowlike nod of his head. I was certainly not offended by either gesture. You may become acclimated to this country more fully after you have been here for awhile, but such gestures are in no way objectionable.

Other body language in greeting or leaving someone can be quite different for different cultures. In the United States, we generally reserve the hug for a rather personal display of affection or a recognition of another person's especially good or bad fortune. For instance, if someone has just won an important award or has just experienced a death in the family, we might give him or her a hug even if ordinarily we would not do so. It seems to be a way to say "congratulations" or "I'm sorry for your misfortune." But hugs, similar to the kiss on the cheek in greetings and leave-taking, are typically reserved for friends and family in most groups in the United States. The handshake is our most common expression as we greet and sometimes leave someone, but it is not always used. Sometimes a touch or pat on the shoulder is an expression used. Whatever your own culture dictates, don't be offended by our shoulder tap or handshake, regardless of the genders involved. While you are in the United States, probably the best idea for any of you who feel unsure about these expressions is to use the simple handshake and talk with American friends about their interpretation of body language and tell them how your culture uses different gestures. Get involved socially with Americans and perhaps join a group in which you will be called upon to listen and talk.

Human beings are somewhat territorial, and the physical distance you stand from someone else is probably a characteristic of the concept of power distance. In the United States, we generally feel comfortable in talking at no closer than arm's length. Power distance has also established precepts for eye contact among individuals in most countries. In the United States, where direct language is preferred and power distance is not great, most people expect direct eye contact between speakers, no matter what position they hold or what age or sex they are. Lack of eye contact often indicates to the American that the speaker is not being truthful or the listener is not listening. So a father may demand of his son, "Look at me when I'm talking to you!" We carry this notion of eye contact into our professional and scientific worlds and expect speakers to make clear eye contact whether the audience is a single individual or a group. Because of this expectation, many international students find communication awkward. Americans also find it awkward to talk with students who resist eye contact. If you detect that a professor is frustrated with you because of lack of eye contact, probably the best thing to do is to explain your background, and the professor will probably understand your ideas of respect. But perhaps try to develop an ability to make that eye contact while you are in the United States. As with other your cultural adaptations, do not lose the original precept; you'll need it when you return to your country. Our method is not better; it is just different.

Even with differences, most body language is fairly easy to understand across cultures. If you detect that a nod of the head or a beckoning with the hand does not fit your meaning for those movements, clarify the issue with the speaker and move on. As with writing styles and speaking patterns with words, try to adapt to the U.S. meanings of body language while you are here. Adaptation is simply increasing your education, not adopting a new culture for yourself. Be proud of yours, and keep it for communication with others in your own culture. The following are my suggestions for the internationals trying to become accustomed to work or study in the United States. If any of them conflict directly with your beliefs or personality, you may have to modify them to satisfy yourself. The first one is most important.

1. Be flexible in your interpretation of us and in your willingness to adjust to this culture.

2. You need not become Americanized; keep what is important to you, but in communicating with American audiences, adjust to their techniques.

3. Listen, read, write, and speak English as much as you can.

4. Get rid of any preconceived ideas you have had about the United States and the people here, and be patient with our attitudes.

5. Be patient with yourself. You have just come into a new world; you will soon adjust to food as well as language.

6. Make friends with Americans; make the first move; start the conversations; and ask for interpretations when meanings are not clear.

7. Observe; watch what respected Americans do and say, and you will soon adapt to what we consider communication norms.

8. Be true to yourself, your personality, and your culture. Adaptation is not adoption.

Of course these are the opinions of an American, but when I asked a colleague of mine who came here some years ago from China what I should tell international students about adjusting to the United States, he said to tell them to be open-minded, patient, flexible, respectful, social, active, precise, right to the point, courteous, truthful, and not too humble or too boastful. He says to observe, ask questions, get involved, and don't isolate yourself. I think he and I are saying essentially the same things.

Most of what we learn about each other's cultures and techniques for communication comes from firsthand experience. The ideas presented here are meant only to stimulate your thinking and understanding as you adjust to communication in a different culture. To assist that understanding, it may help to read the writings of people who have studied these issues in depth. I suggest looking at some Edward T. Hall books, such as *Beyond Culture* (1977), the *Dance of Life* (1983), or Part 1 of *Understanding Cultural Differences* (Hall and Hall, 1990). In that section of the book he and his coauthor, M.R. Hall, discuss key concepts in cultural differences. Especially relative to differences in writing for United States and other cultures, Helen Fox presents very interesting views in *Listening to the World* (1994). Some informative chapters in *Faculty and Student Challenges in Facing Cultural and Linguistic Diversity* (Clark and Waltzman, 1993) deal with students from particular cultures relative to their adjustments in studying in U.S. schools. Montgomery's (2003) suggestions are good for college students, and toward the end of each chapter in *Writing for the Technical Professions*, Woolever (1999) includes a short section on "Tips for International Communication" that may be helpful. Especially if you are reading this book before coming to the United States, I recommend that you also read Robert L. Peters (1997) *Getting What You Came For*, particularly the part in Chapter 23 that is directed to international students.

Whatever our cultural differences and goals of the human spirit, our academic goals and our human motivations and desires are relatively the same worldwide. Our acceptance of those likenesses along with our sincere attempts not only to tolerate but to encourage diversity in communication can lead to greater understanding by both domestic and international students who work together in science, engineering, and technology. Open minds delight in diversity, but for clarity in science we must yield to conventions of the society in which we are communicating. While working in the United States, international students can adapt to U.S. conventions without losing sight of their own personalities, cultures, communication styles, and values. A mark of an educated individual is an ability to communicate with many different audiences in a variety of circumstances.

References

Clark, L. W., and Waltzman, D. E. (1993). *Faculty and Student Challenges in Facing Cultural and Linguistic Diversity*. Charles C Thomas, Springfield, IL.

Day, R. A. (1998). *How to Write and Publish a Scientific Paper*, 5th ed. Oryx Press, Phoenix, AZ.

Fox, H. (1994). *Listening to the World: Cultural Issues in Academic Writing*. National Council of Teachers of English, Urbana, IL.

Gordon, C. H., Simmons, P., Wynn, G., and the Arts Faculty, University of British Columbia. (2003). "Biology Program Guide 2003/2004: Plagiarism." http://www.zoology.ubc.ca/bpg/plagiarism.htm#sel (verified July 2, 2003).

Hall, E. T. (1977). *Beyond Culture*. Anchor Press/Doubleday, Garden City, NY.

Hall, E. T. (1983). *The Dance of Life: The Other Dimension of Time*. Anchor Press/Doubleday, Garden City, NY.

Hall, E. T., and Hall, M. R. (1990). *Understanding Cultural Differences*. Intercultural Press, Yarmouth, ME.

Montgomery, S. L. (2003). *The Chicago Guide to Communicating Science*. University of Chicago Press, Chicago.

Peters, R. L. (1997). *Getting What You Came for: The Smart Student's Guide to Earning a Master's or PhD*, revised ed. The Noonday Press, New York.

Woolever, K. R. (1999). *Writing for the Technical Professions*. Longman, New York.

Appendix 1
WEAKNESSES IN
SCIENTIFIC WRITING

When we mention weaknesses in scientific writing, I'm afraid many people envision misspelled words, too many commas, or sentences that contain other grammatical errors. Those distractions can be serious if they are prevalent in a paper, but they are often the result of larger, more serious issues that the author may have neglected to consider. The most serious problem with most scientific papers is an unwillingness or inability of authors to review their own work objectively and revise it rigorously. Failure to polish a draft is a weakness in itself and allows other weaknesses to be prevalent. An attitude of respect for your audience and sensitivity to language will generally result in clear logic and simplicity, the essentials for good scientific writing. If you feel uncomfortable with detecting weaknesses in your own paper, check carefully on the following elements in your writing. Then read your paper aloud and slowly; you will likely recognize things in that reading that you overlook in silent reading.

LACK OF PREPAREDNESS

Some people begin to write before they are ready or before they have carefully thought through a subject. You must have an audience in mind and have something to say to them. Be confident in your own knowledge and opinions, but to achieve credibility, you must know your subject and believe in what you are saying. You need to support that belief with scientific principles as well as credible information from your own data and the research findings by others. In other words, prepare by studying and understanding the literature, the science, and your data and by talking with others about the subject. As Pauli Wolfgang said, "Ich habe nichts dagegen wenn Sie langsam denken, Herr Doktor, aber ich habe etwas dagegen wenn Sie rascher publizieren als denken (I don't mind if you think slowly, Doctor, but I do mind if you publish faster than you think)." Be prepared before you write and publish.

WEAK ORGANIZATION

A research report must be centered around one point of emphasis. That point, often called the research question or hypothesis, constitutes the purpose of the paper and gives birth to the objectives or supporting points. A logical pattern and progression in arranging ideas in conjunction with that purpose are essential to scientific communication. Almost any scientific report can use the following generic outline as a model, but you must organize details within this pattern.

    I. *Introduction*
      A. *The specific topic and background (literature)*
      B. *Justification and organizational points*
      C. *Objectives of the study*
    II. *Materials and Methods*
      A. *Materials and location of the experiment*
      B. *Procedures and processes*
      C. *Data collection and analysis*
      D. *Statistical evaluations*
    III. *Results and Discussion*
      A. *Synopsis of results*
      B. *Presentation of data (tables, figures, and supporting text)*
      C. *Discussion of significance, application, and relationship to other studies*
    IV. *Conclusions*

Once you have put real content into such a generic outline, the other essential point in organization is **transition** from one idea to another.

INAPPROPRIATE CONTENT
Too Much for One Paper

Excess may come with the paper covering too many diverse points; containing too much material; having too much speculation; being too wordy, redundant, or repetitious; going too far beyond the range of your specific topic; or being overloaded with data that could be expressed in representative samples. Authors can often delete the first paragraph of initial drafts of their papers or reduce it to an introductory sentence. For example, if you are reporting a method for detecting antioxidants in spinach, you need not start by telling the reader that spinach is palatable in salads or as a cooked vegetable. Nor do you need to report historical information on the origin of this species. Get to the point of health benefits from antioxidants from foods such as spinach and methods of analysis. Be concise. As Shakespeare wrote in *King Lear*: "Few words, but to effect."

Too Little in One Paper

Scientists are far more likely to put too much in a paper than too little, but a lack of sufficient development of specific points is also common. Don't presume that your audience will immediately perceive why a statement you make is valid; tell them without being wordy. Writing too much or giving too little content is often the result of being unprepared or not thinking of what the audience needs to know.

POOR STRUCTURE AND UNITY

Closely related to organization, unity within segments (sections, paragraphs, sentences), as well as the transitions between units, is vital to successful construction of the paper (see Chapter 3). Weaknesses in smaller units of the paper occur both in the presentation of data and in sentence construction.

Data

Failure to achieve reader-friendly tables and figures occurs when they contain too little explanation or too many data or meaningless lines. They may be arranged in an unconventional order; have poor or faulty headings, legends, or axis labels; be too complex or cluttered; or be unable to stand alone without the text. Study your own journals and the references listed with Chapter 11 to be certain that your data are presented with as much clarity as possible.

Sentence Construction

Sentences can and should be constructed in a variety of ways. Look at the following sentences; they all say essentially the same thing.

—*Some strata of the earth contain water.*
—*Water is present in some strata of the earth.*

—*Rock and sand strata of the earth may hold water deposits.*
—*Water has been deposited in the earth's strata.*
—*There is water in rock and sand strata of the earth.*
—*Contained within the depths of the earth are extensive strata composed of rock, gravel, or sand, some of which collected large deposits of water billions of years ago and still hold those deposits today.*
—*Deposits called groundwater exist in rock and sand strata of the earth.*
—*Underground are deposits of water.*
—*Hidden in the monstrous recesses of the interior of the earth lie extensive strata of rock and sand wherein there exist enormous volumes of water.*

A dozen other versions of this information could be composed. No single construction is best except in the context in which it appears. Two frequent problems in sentence construction are wordiness and misplaced elements.

Wordiness is the villain that fights the concise, and many of us have trouble recognizing it in our own writing. Just as entire papers often begin with background that is not directly or immediately related to the specific topic of a report, any sentence or paragraph can fail to start with primary words. Look at the following example that came from a graduate student:

> *It can be noted that salmonellae are present during all phases of poultry production and processing. Although similar hygiene practices were practiced on all of the 10 poultry farms we examined in this study, great variation existed in the degree of salmonella contamination on them. From the results of this study, it appears that salmonellae may be transmitted continuously through feed to the breeder parent stock, to the chicks, through the processing and finally to the finished broiler product.*

We can communicate the same message with:

> *Salmonellae is transmitted progressively from feed to breeder chickens and their offspring and then through the processing plant to the finish product. On 10 poultry farms using similar production practices, we found great differences in the degree of salmonella contamination.*

The two ideas in these sentences need further development, probably in separate paragraphs. It may not matter which of the sentences is developed first or last. We can get rid of the repetition of such things as "*in this study,*" "*practices were practiced,*" and "*all phases of poultry production and processing.*" We can also omit rather meaningless phrases such as "*It can be noted that....*" or "*it appears that.*" The result is that we have cut the length almost in half, and our two points are more prominently displayed and ready to be developed with supporting details.

Misplaced elements in sentence construction interfere with smooth reading or logical sequence of ideas. Again, students have furnished examples:

Neither callus tissue from the spinach culture in 1988 nor 1989 produced shoots.

Logically, the neither/nor construction should connect only the years:

Callus tissue from the spinach culture produced shoots in neither 1988 nor 1989.

And another student wrote,

Our purpose was to determine whether the cultivar was more tolerant than others to the pathogen and to characterize the wilt.

Notice how *"to the pathogen"* and *"to characterize the wilt"* appear almost parallel until we give them a second thought. Yes, intelligent readers can quickly understand your meaning, but don't require that they pause or reread a sentence if you can make it flow more simply. Actually, the parallel elements are *"to determine"* and *"to characterize,"* and the sentence reads more smoothly if we put the shorter element closer to the verb and write:

Our purposes were to characterize the wilt and to determine whether the cultivar was more tolerant than others to the pathogen.

DISTRACTING LITTLE THINGS

The reader's attention is distracted by small inconsistencies and errors. Be consistent, and follow a style sheet. Is it Figure, figure, Fig, fig, Fig., fig., or *Figure, figure, Fig, fig, Fig.,* or *fig.* The term might even be written in other ways with underlining or boldface. Hyphens and numbers are used correctly in a variety of ways: for example, *a 3-cm depth, 3 cm deep, two rates, a 2-mg rate, six plants,* or *42 plants.* Decide which forms to use relative to your publisher's style, and use little elements such as abbreviations, hyphens, numbers, and capitalization accurately and consistently.

Errors and inconsistencies in bibliographies and textual citations are distracting and may cause a reader to fail to find a reference. Follow publishers' styles and grammatical standards, and be sure textual citations coordinate with the entries in the list of references. Misspellings, faulty punctuation, and grammatical errors or inconsistencies anywhere in the text or bibliography can frustrate a reader.

SENSITIVITY TO WORDS (DICTION)

Watch out for words with similar meanings or forms. Use the best one available. Sometimes you can't trust the definitions or synonyms as presented in a dictionary or thesaurus. You have to understand conventional use of terms and idioms. For example in scientific papers, we can hardly equate the word *significant* with *important* because of the claim that statistics has laid on *significant*. Dangers exist in accuracy and connotations relative to words such as *affect, effect; medium, media; data; different, varying; while; only; cheap, economical, inexpensive; there is;* and *a small size or a red color.* Avoid the use of groups of words that express the same meaning as a single word. For example, *at this point in time* means *now,* and *a long time period* means *a long time.* Proofread carefully, and don't trust a computer spell-check entirely. Be sensitive to spelling and meanings of words. Believe it or not, the following have appeared in papers I have reviewed. They were probably all spell-checked.

—*This system has been wildly applied by most laboratories in the United States...*
—*The amount of plant material...was not sadistically different between years...*
—*Cotton responds to both soil moisture and relative humility.*
—*Magnesium sulfate had a notorious decrease on the value of moisture absorbed.*

Reading your work aloud and slowly will often call your attention to such oversights and make you smile before an audience laughs out loud.

Appendix 2

THE FIRST DRAFT

The following simple, rough essay was written as a result of a challenge from some undergraduates who assured me that no one can write an answer to an essay question in 10 or 15 minutes. I limited myself to 10 minutes. I chose a subject that I know something about, but we assume that you know your subjects. My answer is crude and far from perfect. It ends abruptly, and I can think of many points that I could have used better. If I were writing a real paper on this subject, this draft would simply provide a beginning. See the discussion and language analysis that follow the essay.

WHAT POINTS DISTINGUISH A GOOD WESTERN SADDLE FROM A POOR ONE?

Materials and workmanship distinguish a good saddle from a bad one. A good saddle will be made on a quality tree with good leather. The saddle tree is the form on which the saddle is built. No saddle is stronger than the foundation. Rawhide-covered wood and molded fiberglass are the strongest trees on the market today. Canvas-covered trees or those with rawhide bindings should be avoided. The leather should be thick and flexible, especially at points of strain (stirrup leathers, rigging, etc.). Thin leather that is heavily hand-tooled is weakened by the tooling. Decorative metal can also weaken leather.

In addition to good materials, the saddle must be put together well. Note whether the leather on the seat is all one piece or is weakened by being sewed together. If possible, check to see how the rigging is attached and be sure that stirrup leathers are one solid piece laced around the tree rail and running the length of the fender. Even check to see where screws, nails, or weaker staples are used.

If the workmanship and materials are satisfactory, buy the saddle on the basis of comfort and application to the job. An uncomfortable saddle is not a good one—sit in it before you buy if you can. If you plan to rope from it, check the horn and swell carefully to see that they will fit that job. A good saddle is one that not only looks good, but is up to a job. (time is up)

Rather than finding fault with this essay, look at what is good about it. Look at the marked copy on the following page. I immediately go to a comparison/contrast and an enumeration approach. The "good and poor" and "points" in the initial question tell me to use these approaches, but notice that I also use definition and even imply cause/effect. Rather than waste time pondering all sorts of possibilities, I start with obvious general points that are important in a saddle (materials and workmanship). Later, I thought of a third important point (the job the saddle is designed to do). I work the third point in as best I can. In this rough draft, I won't worry about the point not being introduced, but in revision I will certainly add it to the two points in the first sentence. From my main points, I went to more specific examples, getting down to tangible pieces of a saddle (tree, rawhide, stirrup leathers). Such elements of support are necessary in a good essay—always! Notice that my concluding remark would have been worthless without the support of the roping saddle. The essay contains good transitions at the beginnings of the second and third paragraphs, and the first sentence in each designates the topic for those paragraphs.

Note also many things that require revision. Sentences lack variety in construction; verbs and other terminology are weak. Ideas need further development. Also, I need to determine whether my audience will understand the saddle jargon. I may have to define more terms. Along the way I shifted from third person to second person. That too should be revised, but I wouldn't worry about it in a first draft. And I will need a better conclusion that reiterates all three of my main points; for example, "...is up to the job and is well crafted from quality materials." Get a rough draft of your paper done with attention to organization and development, and then the real work can begin.

WHAT POINTS DISTINGUISH A GOOD WESTERN SADDLE FROM A POOR ONE?

(enumeration)   *(comparison/contrast)*
Materials and workmanship distinguish a good saddle from a bad one. A
(***Point 1**: topic sentence*)
good saddle will be made on a quality tree with good leather. The saddle tree is
(definition)
the form on which the saddle is built. No saddle is stronger than the foundation.
(enumeration of examples)
Rawhide-covered wood and molded fiberglass are the strongest trees on the
(contrast)
market today. Canvas-covered trees or those with rawhide bindings should be
(enumeration of details)
avoided. The leather should be thick and flexible, especially at points of strain
(tangible examples)   *(contrast)*   *(detail)*
(stirrup leathers, rigging, etc.) Thin leather that is heavily hand-tooled is
(detail)
weakened by the tooling. Decorative metal can also weaken leather.
(transition)   (***Point 2**: topic sentence*)
  In addition to good materials, the saddle must be put together well. Note
(example)   *(contrast)*
whether the leather on the seat is all one piece or is weakened by being sewed
(example)
together. If possible, check to see how the rigging is attached and be sure that
(example)   *(tangible details)*
stirrup leathers are one solid piece laced around the tree rail and running the
(tangible details/examples)
length of the fender. Even check to see where screws, nails, or weaker staples are
used.
(transition)
  If the workmanship and materials are satisfactory, buy the saddle on the
(***Point 3**: topic sentence*)   *(contrast)*
basis of comfort and application to the job. An uncomfortable saddle is not a
(example)
good one — sit in it before you buy if you can. If you plan to rope from it, check
(details)
the horn and swell carefully to see that they will fit that job. A good saddle is one
(generalization)
that not only looks good, but is up to a job. (time is up)

Appendix 3
SAMPLE MANUSCRIPT

The following is an overly simple, abbreviated, fictitious manuscript for publication in a hypothetical journal. Studies that you pursue in science will be much more complex than this study, and your resulting manuscript will also be more complex. You may even find my information questionable. My purpose, however, is not to be scientific but to illustrate the pattern of organization and the techniques for development that should hold true for almost any scientific manuscript. The species, pokeweed, that this study uses is a real plant. People do eat it with preparation to eliminate toxicity. However the study, the data, and the citations are entirely fictitious and serve simply to illustrate my points about organization and development.

OUTLINE

Title: Emergence, Yield, and Quality of Poke Greens from Seeds and Roots

I. Introduction
  A. Definition and research problem
  B. Description of uses
    1. Home preparation
    2. Commercial use
  C. Problems with propagation
    1. Seeds
      a. Germination
      b. Preconditioning

2. *Roots*
   a. *Habitat*
   b. *Handling*
D. *Purpose and objectives*
II. *Materials and Methods*
  A. *Materials*
    1. *Seeds*
      a. *Source*
      b. *Storage*
    2. *Roots*
      a. *Source*
      b. *Storage*
  B. *Procedures*
    1. *Treatments*
      a. *Seeds*
        (1) *Hot water*
          (a) *Time*
          (b) *Temperature*
        (2) *Sulfuric acid*
          (a) *Time*
          (b) *Concentration*
        (3) *Controls*
      b. *Roots*
        (1) *Size*
        (2) *Weight*
    2. *Plantings*
      a. *Seeds*
      b. *Roots*
  C. *Data collection*
    1. *Emergence*
    2. *Dimensions: height, weight, stem diameter*
  D. *Statistical analysis*
III. *Results and Discussion*
  A. *General outcomes*
  B. *Specific outcomes*
    1. *Emergence*
      a. *Roots*
      b. *Seeds*
    2. *Quality comparison*
IV. *Conclusion*

MANUSCRIPT

Emergence, Yield, and Quality of Poke Greens
from Seeds and Roots

Abstract: As commercial value of poke greens increases, so does the need for methods of propagating pokeweed (*Phytolacca americana* L.). We evaluated roots and seeds relative to emergence of shoots and yield and quality of greens produced. Roots were stored in peat moss until planted in three 4-m rows 30 cm apart. Seeds treated with sulfuric acid or hot water soakings were planted in three 3-m rows, and seedlings were thinned to 20 cm after emergence. Untreated seeds served as controls, but emergence was negligible, and data were not useful. Shoots emerged from 87% of the roots and from 67% and 32% of seeds from acid and water soakings, respectively. We harvested 10 randomly selected plants from each treatment 2 weeks after emergence. Seedlings and shoots from roots were not significantly different in height, but shoots from roots had thicker stems, and 10 shoots from roots weighed significantly more at 120 g than did 10 seedlings at 70 g. Roots may be suitable for producing commercial greens, and acid-scarified seeds might produce roots for transplanting.

INTRODUCTION

Pokeweed (*Phytolacca americana* L.), also called poke salad or poke greens, is an edible, perennial herb that reproduces by seeds or roots but is difficult to propagate for commercial use. Although uncooked it is toxic to humans, the young shoots and leaves are often parboiled and washed to remove the toxicity and then cooked for greens, the appearance and taste of which compare with those qualities in spinach. Canners suggest that the greens could be a nutritious and marketable alternative to other greens, but adaptation of the species from the wild to commercial growing conditions has been largely unsuccessful. If propagation methods can be found that produce quality greens, these could fill a timing niche for canning very early in the spring before other crops are ready to process.

    Despite the extra parboiling step in processing, the popularity of poke greens is rather widespread (Smith, 1992), and canners are eager to explore the market. According to S & A Canning Company (Ft. Smith, Arkansas), canning companies in Northwest Arkansas find the commodity profitable enough to pay premium prices for fresh poke greens delivered to their plants on days specified for their canning. The product is shipped to several sites in the United States, especially to California and the Pacific Northwest. The S & A Canning Company processed more than 35,000 cans of poke greens last spring that stayed on grocery shelves an average of only 3 weeks, and the company would like to be able to expand this market (personal communication, Jim Simon, S & A Canning Company). Acquiring enough poke greens for canning by gathering them from native habitats is unlikely to satisfy the demand.

This obstacle could be overcome if the species could be propagated and grown domestically. Until now, propagation and cultivation have been impractical because of poor germination of the seeds and the difficulty in acquiring roots and keeping them viable for planting. Under natural conditions, seeds are seldom viable until they have been digested by birds. In 1972, Evans hypothesized that the seeds that germinate in the wild might be preconditioned in digestive tracts of birds. He collected feces of caged English sparrows fed poke berries and found that seeds from these feces germinated 73% compared with 2% germination for untreated seeds. Lanier and Sizemore (1985) treated pokeweed seed in a hot water bath and achieved 27% germination. Further work on preconditioning seeds might reveal times and temperatures for water or acid treatment that could be even more effective. However, when they are tender enough for greens, shoots produced from seed are typically small with narrow stems (Evans, 1972).

Stems from roots are broader and more succulent, but acquiring the roots is also difficult. Pokeweed often grows in locations that are difficult to reach, and the taproots are very large and lie deep in the soil and subsoil. With a backhoe, Jones et al. (1986) acquired tap roots that measured up to 23 cm wide and 35 cm long. The tops of them were typically located about 8 to 14 cm below the soil surface, and the full root extended as much as 48 cm deep. The roots are fleshy and dehydrate quickly if left exposed to the atmosphere (Evans, 1972). Although far more difficult than seed to acquire and preserve, roots typically produce shoots with thick fleshy stems and relatively large leaves compared with those grown from seeds. Quality of the canned product is superior if at least half the greens are grown from roots (Smith, 1992).

The purpose of our study was to determine the feasibility of propagating and producing quality greens from pokeweed to contribute to an expanded market for the greens. Our objectives were (1) to evaluate germination of pokeweed seeds treated with hot water and sulfuric acid, (2) to determine the feasibility of acquiring and storing roots to successfully grow shoots from them, and (3) to assess the quality of greens produced from both seeds and roots.

MATERIALS AND METHODS

The experiments were conducted in the field at the Western Arkansas Agricultural Experiment Station near Mena in spring of 2001 and 2002. Seeds and roots were acquired from an abandoned field on the Rex Mofield farm near the station. Fully ripened berries were gathered in October of the year before planting and stored at 5°C. Roots were excavated with a backhoe in January of each year, packed in peat moss, and stored at 5°C.

Treatments

Seeds from berries, equilibrated to room temperature overnight, were treated with hot water or sulfuric acid. In a hot water bath, seeds soaked for 1.5 h at temperature maintained at 80°C or for 8 h at 60°C. For the sulfuric acid treatments, we used 10% sulfuric acid in tap water held at 25°C. Seeds were soaked in this treatment for 15 or 30 min. Seeds for the controls were held in cold tap water for 30 min just before they were planted.

Roots measured 11 to 21 cm wide and 24 to 32 cm long. Each of the 32 roots used remained surrounded by at least 6 cm of peat moss until removed for immediate planting. For the 2001 test, roots had been stored for 4 weeks; for 2002, 6 weeks. Initial weights, taken before storing and at removal from the peat moss, indicated an average weight loss of less than 10 g per root.

Both seeds and roots were planted on 15 February in 2001 and on 27 February in 2002. The soil is a Bjorn silt loam, and standard soil tests revealed adequate fertility for optimum pokeweed growth, as determined by the method of Hurter and Balou (1993). The plots had been disked the previous fall, and during the experiment, they were weeded of all species except the pokeweeds. Rainfall supplied adequate moisture for growth.

For each seed treatment and the control, seeds were planted about 2 cm deep and 5 cm apart in three 3-m rows, with each row considered a replication. Rows were 0.5 m apart. Eight days after first emergence, they were thinned to one plant per 20 cm, and any plants that emerged thereafter were rogued. Three rows for roots were 4 m long, and roots were planted 30 cm apart, with the top of the root 10 cm deep in the soil.

Percentage emergence for both seeds and roots was determined at 24 days after planting. Most shoots and seedlings had emerged within 18 days. Plants grew for an additional 2 weeks, at which time 10 plants randomly selected from each treatment were measured for height from the soil surface and cut 2 cm above soil surface to determine weights. Stem diameters were measured at that cut.

Because emergence from the controls was negligible, with only one emerging in 2001 and three in 2002, these were disregarded and comparisons made between roots and seed treatments. Treatments were compared by analyzing data for percentage emergence and plant height, weight, and stem diameter with least significant difference at the 0.05 level.

RESULTS AND DISCUSSION

Percentage emergence from roots was greater than that from seeds, and quality from roots, as determined by weight and stem diameter, was superior to that of seedlings. For the data combined over the 2 years, shoots from roots emerged 87%, with only four of the 32 roots failing to produce shoots. Storage in the peat moss apparently preserved their viability. Jones et al. (1986) found that only two of eight roots kept for 3 weeks in a wooden basket at room temperature produced shoots. Of their two viable roots, one produced two shoots and the other produced one. Of the 28 roots in our study that did develop shoots, 17 produced three to five shoots, 7 produced two shoots, and the other 4 had one each. Size of the root appeared to have no influence on the number of shoots or the size at harvest.

Emergence from seeds treated with sulfuric acid was significantly greater than those from hot water baths (Table 1). Averaged over the 2 years, data showed 73% emergence for the sulfuric acid treatment for 15 min (0.25 h) compared with significantly less at 61% for the 30-min (0.50 h) treatment. This difference might indicate that the acid produced a deleterious effect with time, and further testing might find an even more optimal time than 15 min. Hot water baths did not produce results significantly different from each other, but seeds from either emerged significantly less than from acid treatments with 29% from the 1.5-h soaking and 34% from the 8-h soaking. The somewhat greater emergence from the hot water baths in 2002 than in 2001 may be attributed to weather conditions and the seeds being planted later in 2002.

Heights of seedlings and of shoots from roots were not significantly different, but weights and stem diameters were decidedly different (Table 2).

TABLE 1

Percentage Emergence of Pokeweed from Seeds and Roots from 2 years

Source Treatment (Hours/°C)	Emergence (%)		
	2001	2002	Mean[1]
Roots	83	91	87a
Seeds			
Hot water			
1.5/80	20	38	29c
8.0/60	24	44	34c
Sulfuric acid			
0.25/25	68	78	73b
0.50/25	57	65	61b

[1]Means followed by the same letter are not significantly different at 0.05.

TABLE 2
Two-year Average Yield by Weight, Height, and Stem Diameter of
Pokegreen Plants Grown from Roots and Seeds (2-year Averages
of 10 Plants per Source per Year)

Source	Weight (g)	Height (cm)	Stem diameter (cm)
Roots	120a[1]	16.7a	1.3a
Seeds	70b	18.2a	0.4b

[1]Means followed by the same letter are not significantly different at 0.05.

Shoots from roots were obviously more fleshy at 120 g per 10 shoots compared with the 70 g for 10 seedlings, and they adhered more fully to Smith's (1992) description of preferred greens for canning.

CONCLUSIONS

Our research indicated that the roots were kept viable for transplanting with storage in peat moss, and that acid scarification significantly increased germination of seeds. The higher quality of shoots produced from roots suggests that greens produced perennially from an established field of roots would be preferred over greens grown from seedlings. Acquisition of roots from the wild is still a major hurdle for commercial production. Acid-scarified seeds might be used to establish beds in which to develop roots for transplanting. Further research is needed to establish optimal treatments for seeds and to determine the longevity of production from initially transplanted roots. An expanding market for the canned product may make such research practicable.

References

Evans, D. M. (1972). Germination of pokeweed seeds after scarification in digestive tracts of birds. HortReport 26, 13–16.
Hurter, J. S., and Balou, B. C. (1993). Fertilization and irrigation for rapid growth of poke greens. Sci. Hort. 8, 54–58.
Jones, F. D., Sims, R. T., and Fuller, B. R. (1986). Propagating poke greens from roots. J. Plant Adaptation 13, 65–67.
Lanier, M. H., and Sizemore, Z. T. (1985). Viability of Phytolaca americana with hot water treatment. HortReport 40, 17–20.
Smith, C. A. (1992). Let's capture poke greens. Canners' News 53, 2–3.

Appendix 4
SAMPLE LITERATURE REVIEW

The following literature review was written in conjunction with the proposal in Appendix 5. As any good literature review, it serves (1) as an introduction to the topic, (2) as background information to make the topic more meaningful to the researcher as well as the reader, and (3) as justification for a study. Similar to the proposed study itself, it introduces the problem (polycyclic aromatic hydrocarbon contamination) and possible solutions to the problem (dissipation, remediation), with emphasis on the solution proposed in the study (phytoremediation with microorganisms in the rhizosphere). Organization and content for the review are focused around these key issues, which can be used as the key words for the literature search that provided the content. Note the review is not just an annotated listing of reports published in the area but a smooth-flowing discussion of the issues with appropriate references interspersed. See how Mr. Gentry moved from the general subject to those areas specifically related to his own and ends with a concluding justification for his own study. Also note how he has used headings and transitions to lead the reader through the review. (This review of literature, the proposal in Appendix 5, and an accompanying slide set depicted in Appendix 13 are the work of Terry J. Gentry, a master's candidate at the time he produced this work, now Dr. Terry J. Gentry with Oak Ridge National Laboratory in Oak Ridge, TN. All are used with his permission.)

POLYCYCLIC AROMATIC HYDROCARBON INFLUENCE ON RHIZOSPHERE MICROBIAL ECOLOGY

Contamination of soil by toxic organic chemicals is widespread and frequent. This is sometimes the result of large-scale incidents such as the Exxon Valdez oil spill in Alaska (Pritchard and Costa, 1991). But, more often, smaller areas of soil are polluted. Cole (1992) estimated that in the United States there are 0.5 to 1.5 million underground storage tanks leaking into the surrounding soil. *In situ* bioremediation of these contaminated sites may be more feasible than chemical and physical clean-up methods. Degradation of polycyclic aromatic hydrocarbons (PAHs), a major constituent of many of these pollutants, can be possible if PAH-degrading microorganisms are present at the site. These microorganisms may be more prolific in the rhizosphere of plants than in soil with no vegetation.

A. Polycyclic Aromatic Hydrocarbons (PAHs)

Polycyclic aromatic hydrocarbons are organic compounds that are typically toxic and recalcitrant (Sims and Overcash, 1983). They consist of at least three benzene rings joined in a linear, angular, or cluster array (Cerniglia, 1992). Edwards (1983) described PAHs as being practically insoluble in water. They are produced by various processes, including the incomplete combustion of organic compounds such as petroleum (Giger and Blumer, 1974; Laflamme and Hites, 1978). The carcinogenicity of many PAHs has been well documented (Haddow, 1974). This knowledge has prompted much research to determine the mode by which these compounds cause cancer and their ultimate health risks to humans (Miller and Miller, 1981). Due to their toxic nature, the United States Environmental Protection Agency included several PAHs in their list of priority pollutants to be monitored in industrial wastewaters (Keith and Telliard, 1979). Heitkamp and Cerniglia (1988) concluded that this interest has resulted in increased efforts to remediate PAH-contaminated soil.

B. Dissipation

Reilley et al. (1996) reported the fate of PAHs in soil includes irreversible sorption, leaching, accumulation by plants, and biodegradation. They also contended that surface adsorption is the main process controlling PAH destination in soil. Many PAHs are strongly adsorbed to soil particles (Knox et al., 1993). Means et al. (1980) found the PAHs composed of longer chains and greater masses to be more strongly adsorbed to soil particulate matter. Leaching of PAHs from soil is minimal due their adsorption to soil particles and low water solubility (Reilley et al., 1996). Results indicate that larger PAHs may adsorb onto roots, but translocation from roots to foliar portions of the plants is negligible (Edwards, 1983; Sims and Overcash, 1983). Biodegradation is the main pathway by which dissipation can be enhanced.

C. Bioremediation

Bioremediation manipulates biodegradation processes by using living organisms to reduce or eliminate hazards resulting from accumulation of toxic chemicals and other hazardous wastes. According to Bollag and Bollag (1995), two techniques that may be used in bioremediation are (1) stimulation of the activity of the indigenous microorganisms by the addition of nutrients, regulation of redox conditions, optimization of pH, or augmentation of other conditions to produce an environment more conducive to microbial growth and (2) inoculation of the contaminated sites with microorganisms of specific biotransforming abilities.

1. Indigenous Population. Soil contains a large and diverse population of microorganisms (Tiedje, 1994). The indigenous population of these microorganisms has been manipulated to increase biodegradation. *In situ* bioremediation utilizes organisms at the site of pollution to remove contaminants. Often, indigenous organisms from the contaminated area, which may even have adapted to proliferate on the chemical, are utilized to remove the pollutants (Bollag and Bollag, 1995).

Microbial degradation may be enhanced by aeration, irrigation, and application of fertilizers (Lehtomäki and Niemelä, 1975). In Prince William Sound, Alaska, following the Exxon Valdez oil spill, the application of fertilizers increased biodegradation up to threefold (Pritchard and Costa, 1991).

The relative contributions of bacterial and fungal populations to hydrocarbon mineralization may differ based upon contaminant and soil parameters. Anderson and Domsch (1975) studied the degradation of glucose in several soils. They attributed the majority of mineralization to fungi (60%–90%) with relatively minor bacterial contribution (10%–40%). It is unclear if fungi are also the principal degraders of hydrocarbons (Bossert and Bartha, 1984). Song et al. (1986) reported 82% of n-hexadecane mineralization was due to bacteria while fungi contributed only 13%. They concluded that bacteria are the primary degraders of n-hexadecane in the soil tested, but additional experiments are necessary before the results can be generalized. In a field study utilizing six oils as contaminants, Raymond et al. (1976) noted that fungi appeared to be the principal hydrocarbon-degraders.

From a review of the literature, Cerniglia (1992) found various bacteria, fungi, and algae reported to degrade PAHs. More specifically, Déziel et al. (1996) isolated 23 bacteria capable of utilizing naphthalene and phenanthrene as their sole growth substrate. These bacteria were all fluorescent pseudomonads. Shabad and Cohan (1972) reported that soil bacteria are the primary degraders of PAHs. Cerniglia's (1992) review concluded that the microbial degradation of smaller PAHs such as phenanthrene has been thoroughly investigated; however, there has not been sufficient research on the microorganisms capable of degrading PAHs containing four or more aromatic rings. There remains a need for isolation and identification of microorganisms capable of degrading the more persistent and toxic PAHs (Cerniglia, 1992).

2. Introduced Microorganisms. Organisms capable of breaking down certain pollutants are not present at all sites; therefore, inoculation of the soil with microorganisms, or bioaugmentation, is sometimes required (Alexander, 1994). Indigenous or exogenous microorganisms may be applied to the polluted soil (Turco and Sadowsky, 1995). Microorganisms capable of degrading several pollutants including PAHs have been isolated from contaminated soil (Heitkamp and Cerniglia, 1988). In addition, Lindow et al. (1989) communicated a need for the continued development of genetically engineered microorganisms including those capable of degrading a variety of pollutants. Nevertheless, successful establishment of introduced microorganisms remains enigmatic (Pritchard, 1992; Turco and Sadowsky, 1995). Thies et al. (1991) linked the poor survival of introduced microorganisms to competition from native soil microorganisms.

The characteristics that allow introduced microorganisms to become acclimated to a new environment have not been completely elucidated (Turco and Sadowsky, 1995). However, the indigenous soil populations appear to have specific qualities, such as the ability to utilize a particular growth substrate, that give them a competitive advantage in occupying available niches (Atlas and Bartha, 1993). One way to encourage the growth of introduced microorganisms may be to supply a new niche for microbial growth in the form of a suitable plant.

D. Phytoremediation

Phytoremediation is defined by Cunningham and Lee (1995) as "the use of green plants to remove, contain, or render harmless environmental contaminants." This applies to all plant-influenced biological, microbial, chemical, and physical processes that contribute to the remediation of contaminated sites (Cunningham and Berti, 1993). Plants have historically been developed for food or fiber production. With an increasing interest in the use of plants to reduce contamination from organic chemicals, plants may be selected and developed based upon their suitability for bioremediation. Cunningham and Lee (1995) contend that plant attributes such as rooting depth, structure and density can be altered to increase biodegradation. They assert that, if contaminants are (1) in the upper portion of the soil, (2) resistant to leaching, and (3) not an immediate hazard, many may be removed by phytoremediation. Experiments may confirm that phytoremediation is a less expensive, more permanent, and less invasive technique than many current methods of remediation (Cunningham and Lee, 1995).

E. The Rhizosphere

Curl and Truelove (1986) have described the rhizosphere as the zone of soil under the direct influence of plant roots and in which there is an increased level of microbial numbers and activity. They report that the ratio of bacteria and fungi in the rhizosphere to the non-rhizosphere soil (R/S ratio) commonly ranges from 2 to 20 due to the root exudation of easily metabolizable substrates. These exudates include sugars, amino compounds, organic acids, fatty acids, growth factors, and nucleotides (Curl and Truelove, 1986). Legumes usually demonstrate a greater rhizosphere effect than non-legumes (Atlas and Bartha, 1993). Also, the development of plant roots in previously nonvegetated soil may alter soil environmental conditions including carbon dioxide and oxygen concentrations, osmotic and redox potentials, pH, and moisture content (Anderson and Coats, 1995).

Generally, the rhizosphere is colonized by a predominantly gram-negative bacterial community (Curl and Truelove, 1986). Anderson and Coats (1995) reported that one of the interesting and repeated topics discussed during the 1993 American Chemical Society symposium was the prevalence of gram-negative microorganisms in the rhizosphere. Reportedly, the ability of gram-negative bacteria to quickly metabolize root exudates contributes to their predominance in the rhizosphere (Atlas and Bartha, 1993).

Anderson and Coats (1995) suggest that increased rates of contaminant degradation in the rhizosphere compared to nonvegetated soil may result from increased numbers and diversity of microorganisms.

1. Rhizosphere Effect on PAHs. The rhizosphere of numerous plants has been reported to increase the biodegradation of several PAHs. Aprill and Sims (1990) examined the effects of eight prairie grasses (*Andropogon gerardi, Sorghastrum nutans, Panicum virgatum, Elymus canadensis, Schizachyrium scoparius, Bouteloua curtipendula, Agropyron smithii, and Bouteloua gracilis*) on the biodegradation of four PAHs, benzo(a)pyrene, benz(a)anthracene, chrysene, and dibenz(a,h)anthracene. They reported significantly greater disappearance of the PAHs in the vegetated soils compared to the unvegetated soils, and the rate of disappearance was directly related to the water solubility of each compound.

Reilley et al. (1996) investigated the rhizosphere effect of alfalfa (*Medicago sativa* L.), fescue (*Festuca arundinacea* Schreb.), sudangrass (*Sorghum vulgare* L.), and switchgrass (*Panicum virgatum* L.) on the degradation of pyrene and anthracene. They reported that the vegetation significantly increased the degradation of these PAHs in the soil. They concluded that degradation most likely resulted from an elevated microbial population in the rhizosphere due to the presence of root exudates.

Nichols et al. (1996) conducted an experiment on the degradation of a model organic contaminant (MOC) composed of six organic chemicals including two PAHs (phenanthrene and pyrene) in the rhizospheres of alfalfa (*Medicago sativa*, var. Vernal) and alpine bluegrass (*Poa alpina*). They found increased numbers of hydrocarbon-degrading microorganisms in the rhizospheres of both plants.

From the same study, Rogers et al. (1996) reported that plants demonstrated no significant impact on the degradation of the MOC. They concluded it was probable that biological and/or abiotic processes occurring before plants developed enough to produce a rhizosphere effect were responsible for the disappearance of the MOC compounds.

2. Rhizosphere microbial ecology in PAH-contaminated soil. Before microorganisms can be successfully introduced into the soil or managed for increased bioremediation, an increased understanding of the determinants of rhizosphere microbial ecology needs to be developed. Anderson and Coats (1995) stated the need for an expanded understanding of the interactions between plants, microorganisms, and chemicals in the root zone in order to identify conditions where phytoremediation using rhizosphere microorganisms is most feasible.

To date, no studies have been conducted on rhizosphere microbial ecology in PAH-contaminated soil. Furthermore, little is known about the factors controlling rhizosphere microbial ecology in uncontaminated soil. Bowen (1980) asserted the plant to be the predominant force in the rhizosphere system. In contrast, Bachmann and Kinzel (1992) reported that, in a study involving six plants and four soils, the soil was the dominant factor in some plant-soil combinations. A unique symbiosis that developed from the combination of a specific plant and soil microorganisms was evident. Of all tested plants, *Medicago sativa* had the strongest influence on the soil. They concluded that this effect was consistent with the results of Angers and Mehuys (1990) and may be related to the nitrogen-fixing activity of alfalfa.

Additionally, recent research suggests that gram-positive bacteria may be a larger component of the rhizosphere microbial population than previously reported. Cattelan et al. (1995) found a large percentage of soybean (*Glycine max*) rhizosphere population to be occupied by the gram-positive bacterial genus *Bacillus* spp. Also, it appears that gram-positive microorganisms may play a major role in the breakdown of contaminants including PAHs. Heitkamp and Cerniglia (1988) isolated a gram-positive bacterium capable of degrading several PAHs. The bacterium could not utilize PAHs as the sole C source, but it did completely mineralize PAHs when supplied with common organic carbon sources such as peptone and starch. Additional research is needed to elucidate the determinants of rhizosphere microbial ecology especially in PAH-contaminated soils.

References

Alexander, M. (1994). *Biodegradation and Bioremediation*. Academic Press, San Diego, CA.
Anderson, J. P. E., and Domsch, K. H. (1975). Measurement of bacterial and fungal contributions to respiration of selected agricultural and forest soils. *Can. J. Microbiol.* **21**, 314–322.

Anderson, T. A., and Coats, J. R. (1995). An overview of microbial degradation in the rhizosphere and its implications for bioremediation. In *Bioremediation: Science and Applications* (H. D. Skipper and R. F. Turco, eds.), SSSA Spec. Publ. 43, pp. 135–143. ASA, CSSA, and SSSA, Madison, WI.

Angers, D. A., and Mehuys, G. R. (1990). Barley and alfalfa cropping effects on carbohydrate contents of a clay soil and its size fractions. *Soil Biol. Biochem.* **22,** 285–288.

Aprill, W., and Sims, R. C. (1990). Evaluation of the use of prairie grasses for stimulating polycyclic aromatic hydrocarbon treatment in soil. *Chemosphere* **20,** 253–265.

Atlas, R. M., and Bartha, R. (1993). *Microbial Ecology: Fundamentals and Applications,* 3rd ed. Benjamin/Cummings, Menlo Park, CA.

Bachmann, G., and Kinzel, H. (1992). Physiological and ecological aspects of the interactions between plant roots and rhizosphere soil. *Soil Biol. Biochem.* **24,** 543–552.

Bollag, J.-M., and Bollag, W. B. (1995). Soil contamination and the feasibility of biological remediation. In *Bioremediation: Science and Applications* (H. D. Skipper and R. F. Turco, eds.), SSSA Spec. Publ. 43, pp. 1–12. ASA, CSSA, and SSSA, Madison, WI.

Bossert, I., and Bartha, R. (1984). The fate of petroleum in soil ecosystems. In *Petroleum Microbiology* (R. M. Atlas, ed.), pp. 435–473. Macmillan, New York.

Bowen, G. D. (1980). Misconceptions, concepts and approaches in rhizosphere biology. In *Contemporary Microbial Ecology* (D. C. Ellwood, J. N. Hedger, M. F. Latham, J. M. Lynch, and J. H. Slater, eds.), p. 283–304. Academic Press, New York.

Cattelan, A. J., Hartel, P. G., and Fuhrmann, J. J. (1995). Successional changes of bacteria in the rhizosphere of nodulating and non-nodulating soybean (*Glycine max* (L.) Merr.). 15th North American Conference on Symbiotic Nitrogen Fixation, August 13–17, 1995, Raleigh, NC.

Cerniglia, C. E. (1992). Biodegradation of polycyclic aromatic hydrocarbons. *Biodegradation* **3,** 351–368.

Cole, G. M. (1992). *Underground Storage Tank Installation and Management.* Lewis, Chelsea, MI.

Cunningham, S. D., and Berti, W. R. (1993). The remediation of contaminated soils with green plants: An overview. *In Vitro Cell. Dev. Biol.* **29P,** 207–212.

Cunningham, S. D., and Lee, C. R. (1995). Phytoremediation: Plant-based remediation of contaminated soils and sediments. In *Bioremediation: Science and Applications* (H. D. Skipper and R. F. Turco, eds.), SSSA Spec. Publ. 43, pp. 145–156. ASA, CSSA, and SSSA, Madison, WI.

Curl, E. A., and Truelove, B. (1986). *The Rhizosphere.* Springer-Verlag, Berlin.

Déziel, É., Paquette, G., Villemur, R., Lépine, F., and Bisaillon, J.-G. (1996). Biosurfactant production by a soil *Pseudomonas* strain growing on polycyclic aromatic hydrocarbons. *Appl. Environ. Microbiol.* **62,** 1908–1912.

Edwards, N. T. (1983). Polycyclic aromatic hydrocarbons (PAH's) in the terrestrial environment— a review. *J. Environ. Qual.* **12,** 427–441.

Giger, W., and Blumer, M. (1974). Polycyclic aromatic hydrocarbons in the environment: Isolation and characterization by chromatography, visible, ultraviolet, and mass spectrometry. *Anal. Chem.* **46,** 1663–1671.

Haddow, A. (1974). Sir Ernest Laurence Kennaway FRS, 1881–1958: Chemical causation of cancer then and today. *Perspectives Biol. Med.* **17,** 543–588.

Heitkamp, M. A., and Cerniglia, C. E. (1988). Mineralization of polycyclic aromatic hydrocarbons by a bacterium isolated from sediment below an oil field. *Appl. Environ. Microbiol.* **54,** 1612–1614.

Keith, L. H., and Telliard, W. A. (1979). Priority pollutants. I. A perspective view. *Environ. Sci. Technol.* **13,** 416–423.

Knox, R. C., Sabatini, D. A., and Canter, L. W. (1993). *Subsurface Transport and Fate Processes.* Lewis, Boca Raton, FL.

Laflamme, R. E., and Hites, R. A. (1978). The global distribution of polycyclic aromatic hydrocarbons in recent sediments. *Geochim. Cosmochim. Acta* **42,** 289–303.

Lehtomäki, M., and Niemelä, S. (1975). Improving microbial degradation of oil in soil. *Ambio* **4**, 126–129.

Lindow, S.E., Panopoulos, N. J., and McFarland, B. L. (1989). Genetic engineering of bacteria from managed and natural habitats. *Science* **244**, 1300–1307.

Means, J. C., Wood, S. G., Hassett, J. J., and Banwart, W. L. (1980). Sorption of polynuclear aromatic hydrocarbons by sediments and soils. *Environ. Sci. Tech.* **14**, 1524–1528.

Miller, E. C., and Miller, J. A. (1981). Searches for ultimate chemical carcinogens and their reactions with cellular macromolecules. *Cancer* **47**, 2327–2345.

Nichols, T. D., Wolf, D. C., Rogers, H. B., Beyrouty, C. A., and Reynolds, C. M. (1996). Rhizosphere microbial populations in contaminated soils. *Water Air Soil Pollut.* (in press).

Pritchard, P. F., and Costa, C. T. (1991). EPA's Alaska oil spill bioremediation project. *Environ. Sci. Technol.* **25**, 372–379.

Pritchard, P. H. (1992). Use of inoculation in bioremediation. *Current Opin. Biotech.* **3**, 232–243.

Raymond, R. L., Hudson, J. O., and Jamison, V. W. (1976). Oil degradation in soil. *Appl. Environ. Microbiol.* **31**, 522–535.

Reilley, K. A., Banks, M. K., and Schwab, A. P. (1996). Dissipation of polycyclic aromatic hydrocarbons in the rhizosphere. *J. Environ. Qual.* **25**, 212–219.

Rogers, H. B., Beyrouty, C. A., Nichols, T. D., Wolf, D. C., and Reynolds, C. M. (1996). Selection of cold-tolerant plants for growth in soils contaminated with organics. *J. Soil Cont.* **5**, 171–186.

Shabad, L. M., and Cohan, Y. L. (1972). The contents of benzo(a)pyrene in some crops. *Arch. Geschwultsforsch.* **40**, 237–246.

Sims, R. C., and Overcash, M. R. (1983). Fate of polynuclear aromatic compounds (PNA's) in soil-plant systems. *Residue Rev.* **88**, 1–68.

Song, H.-G., Pedersen, T. A., and Bartha, R. (1986). Hydrocarbon mineralization in soil: Relative bacterial and fungal contribution. *Soil Biol. Biochem.* **18**, 109–111.

Thies, J. E., Singleton, P. W., and Bohlool, B. B. (1991). Influence of the size of indigenous rhizobial populations on establishment and symbiotic performance of introduced rhizobia on field-grown legumes. *Appl. Environ. Microbiol.* **57**, 19–28.

Tiedje, J. M. (1994). Microbial diversity: Of value to whom? *Am. Soc. Microbiol.* News **60**, 524–525.

Turco, R. F., and Sadowsky, M. (1995). The microflora of bioremediation. In *Bioremediation: Science and Applications* (H. D. Skipper and R. F. Turco, eds.), SSSA Spec. Publ. 43, pp. 87–102. ASA, CSSA, and SSSA, Madison, WI.

Appendix 5
SAMPLE PROPOSAL

As with the literature review in Appendix 4, the following proposal is the work of Terry J. Gentry and is used with his permission. Both the review of literature and the proposal were presented to his graduate committee along with his proposed plan of course work. Because this proposal stands as a document separate from the Review of Literature, it contains some repetition from that review. An accompanying slide set, copied in Appendix 13, was used with an oral presentation and was also presented to his committee with his proposal. In addition, his preliminary work was used as the basis for a poster at a regional meeting.

Notice in this proposal the inclusion of the preliminary work, which is not always a part of a graduate proposal. The proposal could have been presented to the graduate committee before the preliminary work. In that situation, the proposal would contain two objectives, and conditions of the second objective and methods would depend upon results from the first. With inclusion of the preliminary study here, however, Mr. Gentry has additional justification for his proposed study and has already determined the conditions on which the objective and methods will be based. The methods section of his "Proposed Experiment" can simply refer to that section of the "Preliminary Study." Also, some graduate committees may want the literature review incorporated into the proposal itself. The following model, then, is not to be considered a set standard; individual proposals must fit the situation, the committee requirements, and the experimentation proposed.

POLYCYCLIC AROMATIC HYDROCARBON INFLUENCE ON RHIZOSPHERE MICROBIAL ECOLOGY

INTRODUCTION

Contamination of soil by toxic organic chemicals is widespread and frequent. This is sometimes the result of large-scale incidents such as the Exxon Valdez oil spill in Alaska (Pritchard and Costa, 1991). But, more often, smaller areas of soil are polluted. Cole (1992) estimated that in the United States there are 0.5 to 1.5 million underground storage tanks leaking into the surrounding soil. Polycyclic aromatic hydrocarbons (PAHs) are a major constituent of many of these pollutants. These PAHs are characterized by their toxicity and persistence (Sims and Overcash, 1983).

Phytoremediation is a technology that may provide cheaper, more permanent, and less invasive amelioration of these contaminated soils than other remediation methods (Cunningham and Lee, 1995). Curl and Truelove (1986) described the rhizosphere as the zone of soil under the direct influence of plant roots and in which there is an increased level of microbial numbers and activity. The rhizosphere of numerous plants has been shown to increase the biodegradation of various organic contaminants (Anderson and Coats, 1995). Aprill and Sims (1990) found increased disappearance of four PAHs in soil columns planted with prairie grass compared with nonvegetated soil columns. Reilley et al. (1996) reported an increased degradation of pyrene and anthracene in the rhizospheres of three grasses and a legume.

Although much work has been done on manipulating the soil microflora, little is known about the determinants of rhizosphere microbial ecology (Bachmann and Kinzel, 1992). Anderson and Coats (1995) suggest that a better understanding of the mechanistic interactions among plants, microorganisms, and chemicals in the root zone will help identify situations where phytoremediation can be appropriate.

In a preliminary study we assessed the diversity of microorganisms present in two soils without vegetation and in the rhizosphere of bahiagrass (*Paspalum notatum* Flügge, var. Argentine). The goal of the proposed study is to determine how populations of microorganisms in the rhizosphere respond to the presence of a PAH.

PRELIMINARY STUDY

The objective of the preliminary experiment was to assess the impact of bahiagrass rhizosphere on bacterial density and diversity in two soils.

Materials and Methods

The experiment was conducted under preset conditions on a Captina silt loam (fine-silty, siliceous, mesic Typic Fragiudults) at the University of Arkansas—Fayetteville and on an Appling sandy loam (clayey, kaolinitic, thermic Typic Kanhapludults) at the University of Georgia—Athens. Field moist samples from the Ap horizons of the soils were passed through sterile 2-mm sieves and stored at 4°C until the beginning of the experiment. Cone-tainers® were surface-sterilized by submerging in 30% bleach for 15 min and rinsing with sterile deionized water. Sterile cheesecloth patches were placed in the bottom of Cone-tainers®. Sixty grams of moist soil (52.4 g dry weight) was weighed into respective surface-sterilized Cone-tainers® and incubated in growth chambers with day/night cycles of 16/8 h and 27/16 ± 1°C. For 2 weeks prior to planting, soils were maintained at 60% of –0.03 MPa (14.2% w/w) soil moisture potential with sterile distilled water.

To reduce plant genetic variability, bahiagrass, an apomictic plant, was selected for the experiment. One gram of seed was weighed into a scintillation vial. Ten milliliters of 95% ethanol was added to the seed and vortexed. Ethanol was removed from seed by pipette. The vial containing ethanol-cleaned seed was placed onto ice in an ice bucket. Ten milliliters of concentrated sulfuric acid was added to the vial. The seed were vortexed a few seconds every minute during the 8-min scarification process. Sulfuric acid was drawn off by pipette, and the seed were rinsed seven times with distilled water.

Scarified seed were surface-sterilized with 10 mL of 30% bleach solution added to the vial. Seed were sterilized for 30 min and were vortexed for a few seconds every 5 min. Bleach solution was drawn off, and 10 mL of sterile 0.01 M HCL was added to the vial, vortexed, and allowed to stand for 10 min. A pipette was used to remove the 0.01 M HCL from the vial, and the seed were rinsed with sterile distilled water six times. Sterile seed were inserted between moist sheets of sterile filter paper in Petri dishes and placed in an incubator at 30 ± 1°C. After 4 days, five germinated seeds were planted in respective, pre-incubated Cone-tainers® and covered with 1 cm of sterile sand. Cone-tainers® were returned to the growth chambers and maintained at the same conditions as prior to planting. Seedlings were thinned to one per Cone-tainer® after emergence. Bulk soil in control Cone-tainers® received no seed but was otherwise treated the same.

Plants were harvested 3 weeks after planting. Cone-tainers® were sectioned vertically with a sterile scalpel, and plant shoots were cut off at the soil surface. Soil was emptied onto sterile aluminum foil. Roots were carefully removed from the soil and lightly shaken against a sterile surface to remove loosely attached soil. Roots plus tightly adhering, rhizosphere soil were placed into 99-mL dilution blanks for the 10^{-2} dilutions. One gram of soil from the non-vegetated soil Cone-tainers® was utilized for the corresponding 10^{-2} dilutions of nonrhizosphere (bulk) soil. Samples were serially diluted and spread-plated on 1/10-strength trypticase soy agar (0.1× TSA) containing 100 mg cycloheximide/L. Total numbers of aerobic, heterotrophic bacteria were determined after incubation of plates at 28 ± 1°C for 2 days (Wollum, 1994).

Approximately 60 well-separated bacterial colonies were randomly selected from each treatment: Captina bulk soil, Captina rhizosphere, Appling bulk soil, and Appling rhizosphere. Isolates were restreaked twice on 0.1× TSA plates to check for purity before a final transfer to 1× BBL brand TSA (Becton Dickinson and Co., Cockeysville, MD). After 24 ± 2 h of growth at 28 ± 1°C, about 40 mg of each isolate was extracted for fatty acid methyl ester (FAME) analysis (Sasser, 1990).

Extracts were sent to the University of Delaware and analyzed by gas chromatography with the MIDI system (MIDI, Newark, DE). Fatty acid profiles were compared with aerobic bacteria in the TSBA library from MIS Standard Libraries. A similarity index of ≥0.4 was considered a match for the genus (Kennedy, 1994; Sasser, 1990).

A 2-sample t-test was used to compare the total bacterial numbers in the bulk soil and rhizosphere for each soil. The distributions of bacterial genera for soil (Captina, Appling)—treatment (bulk soil, rhizosphere) combinations were compared by a chi-squared test for equality of distributions. Where small expected cell counts invalidated the usual chi-squared test, Fisher's exact test was used. The P values at ≤0.05 were considered significant.

Results and Discussion

Total numbers of soil bacteria for the bulk soil and bahiagrass rhizosphere of the Captina silt loam were 1.0×10^7 and 1.1×10^7 colony-forming units (CFU)/g of dry soil, respectively, and were not significantly different (Table 1). Similar data

TABLE 1
Total Bacterial Numbers in Bulk Soil and Bahiagrass Rhizosphere (Rhiz) of Captina and Appling Soils

Captina silt loam			Appling sandy loam		
Bulk	Rhiz	R/S*	Bulk	Rhiz	R/S
-10^6 CFU/g dry soil–			-10^6 CFU/g dry soil–		
10.0 a**	11.0 a	1.1	5.8 a	18.3 b	3.2

*Ratio of rhizosphere/bulk soil populations.
**For a given soil, numbers with the same lowercase letter are not significantly different at the 5% level.

were recently reported for bulk soil and alpine bluegrass (*Poa alpinus*) rhizo-sphere of Captina silt loam (Nichols et al., 1996). The bluegrass demonstrated a slightly larger rhizosphere influence than the bahiagrass on the total number of bacteria, but this difference may have been partly due to the longer growth period utilized in that experiment.

In contrast, total bacterial numbers of the Appling sandy loam were 0.6×10^7 and 1.8×10^7 CFU/g dry soil in the bulk soil and rhizosphere, respectively, and were significantly different. A comparable increase in bacterial numbers has been shown for a soybean (*Glycine max*) rhizosphere in Appling sandy loam (Cattelan et al., 1995).

The observed rhizosphere influence on bacterial numbers was relatively small with R/S ratios of 1.1 for the Captina silt loam and 3.2 for the Appling sandy loam. Published data indicate that grasses typically exhibit low R/S ratios as compared to legumes (Curl and Truelove, 1986).

Although there was no measurable difference in the total number of bacteria between the bulk soil and bahiagrass rhizosphere of Captina silt loam, there was a significant increase in diversity in the rhizosphere (Fig. 1). The percentage of *Bacillus* spp. decreased from 84% in the bulk soil to 25% in the rhizosphere. Accordingly, the number of bacterial genera represented in the rhizosphere was increased with the appearance of *Alcaligenes* and *Burkholderia* + *Pseudomonas* spp. There was a concomitant increase in the proportion of *Arthrobacter* and *Micrococcus* spp. Also, the number of unidentified bacteria (similarity indices <0.40) increased from 12% in the bulk soil to 39% in the rhizosphere.

Despite the decrease in the percentage of *Bacillus* spp., the gram-positive bacteria *Bacillus* and *Arthrobacter* were dominant in the Captina bulk soil and rhizosphere. Only small numbers of gram-negative bacteria such as *Pseudomonas* were identified. These results differ from those reported by other investigators that gram-negative bacteria predominate in the rhizosphere (Curl and Truelove, 1986). The gram status of unidentified isolates has not been determined.

In contrast, Appling sandy loam demonstrated no significant change in bacterial diversity between the bulk soil and rhizosphere (Fig. 2). Only three bacterial genera were identified in the bulk soil and in the rhizosphere. Identified bacteria were mostly *Bacillus* and *Arthrobacter* spp., and their num-bers were relatively consistent at 21% and 10% in the bulk soil and 25% and 10% in the rhizosphere, respectively. The majority of the isolates from the bulk soil and rhizosphere were not identified, 68% and 63%, respectively. Cattelan et al. (1995) reported that there was no significant difference in the relative frequency of bacterial genera of the bulk soil and soybean rhizosphere of Appling sandy loam when sampled at 3 and 15 days after planting.

Again, in contrast to the reported data (Curl and Truelove, 1986), the gram-positive genera *Bacillus* and *Arthrobacter* comprised the majority of identified isolates from the bulk soil and rhizosphere of the Appling sandy loam. In addition,

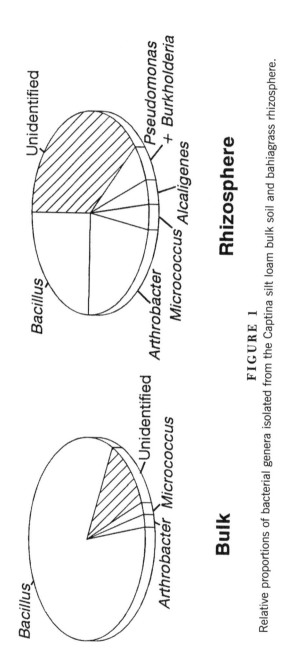

Bulk

Rhizosphere

FIGURE 1

Relative proportions of bacterial genera isolated from the Captina silt loam bulk soil and bahiagrass rhizosphere.

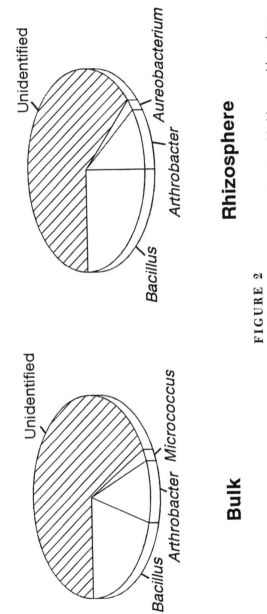

FIGURE 2

Relative proportions of bacterial genera isolated from the Appling sandy loam bulk soil and bahiagrass rhizosphere.

the gram status of a large, unidentified percentage of bacterial isolates has not been determined.

Conclusions

The data indicate that bahiagrass rhizosphere significantly increased the bacterial diversity in the Captina silt loam but did not influence the bacterial diversity in the Appling sandy loam. The soil appeared to impact the rhizosphere effect on bacterial diversity. More soils need to be tested to clarify these results. The total rhizosphere influence on bacterial numbers and diversity may be underestimated due to the inability to culture and/or identify a large portion of the soil bacterial population by current methods.

PROPOSED EXPERIMENT

With some knowledge of its microbial ecology from the preliminary experiment, Captina silt loam has been chosen as the soil to assess the influence of a PAH on the microbial ecology in the rhizosphere. Likewise, bahiagrass will be one of the plants studied. Alfalfa (*Medicago sativa,* var. Vernal), a legume, has been added to the study because of its demonstrated rhizosphere influence (Nichols et al., 1996). Pyrene, a four-ring PAH, will be the contaminant due to its recalcitrance in soil (Rogers et al., 1996).

Objective

The objective of this study is to determine the effects of pyrene on the rhizosphere microbial ecology of bahiagrass and alfalfa.

Materials and Methods

Treatments will be similar to those used in the preliminary study, and comparisons will be made between bulk soil, bahiagrass rhizosphere, and alfalfa rhizosphere with and without pyrene. The Captina silt loam will be collected and sieved through a sterile 2-mm sieve. Sieved soil will be stored at 4°C in Ziploc® bags until the experiment begins. Soil equivalent to 750 g dry weight will be added to glass 1.5 pint jars, which have been sterilized by rinsing once with 30% bleach solution and six times with water to remove residual bleach before soil is added.

Pyrene will be ground in a mortar and pestle, passed through a 250-μm sieve, and mixed into soil at a level of either 0 mg/kg or 2000 mg/kg. Controls with no pyrene will be used. Soil in jars will be pre-incubated for 2 weeks at room temperature. Soil is to be maintained at −0.03 MPa soil moisture potential by daily addition of sterile distilled water.

Bahiagrass and alfalfa seed will be germinated and planted into respective jars using procedures outlined in the preliminary study. Plants will be harvested 10 weeks after planting following procedures outlined in the preliminary study. Samples will also be spread-plated onto Martin's medium to determine the number of fungi and onto starch-casein agar to determine the number of actinomycetes (Wollum, 1994). Approximately 500 colonies each of the actinomycetes, bacteria, and fungi will be isolated from spread-plates and inoculated into Bushnell-Haas broth containing 50 mg/10 mL pyrene to determine whether the microbial isolates are capable of using pyrene as a sole C source and can degrade the compound. Pyrene in the rhizosphere and nonrhizosphere soil will be extracted and respective levels determined by gas chromatographic analysis. Statistical analysis will be the same as those outlined in the preliminary study.

Conclusions

Much work has been done to determine the rhizosphere effect on the biodegradation of organic contaminants. Studies have attributed the increased degradation of several compounds in the rhizosphere of numerous plants to microorganisms. In contrast, very few studies have examined the effect of contaminants on the microbial ecology of the rhizosphere. An increased understanding of the microbial interactions in the rhizosphere may make possible more precise manipulation of the rhizosphere microbial population for the biodegradation of organic contaminants. These experiments should identify specific microorganisms capable of degrading the PAH pyrene. Additionally, the increased knowledge of rhizosphere microbial ecology in contaminated soil may lead to the selection of appropriate plants capable of stimulating an indigenous or introduced microbial population to enhance the remediation of contaminated sites.

References

Anderson, T. A., and Coats, J. R. (1995). An overview of microbial degradation in the rhizosphere and its implications for bioremediation. In *Bioremediation: Science and Applications* (H. D. Skipper and R. F. Turco, eds.), SSSA Spec. Publ. 43, pp. 135–143. ASA, CSSA, and SSSA, Madison, WI.

Aprill, W., and Sims, R. C. (1990). Evaluation of the use of prairie grasses for stimulating polycyclic aromatic hydrocarbon treatment in soil. *Chemosphere* **20**, 253–265.

Bachmann, G., and Kinzel, H. (1992). Physiological and ecological aspects of the interactions between plant roots and rhizosphere soil. *Soil Biol. Biochem.* **24**, 543–552.

Cattelan, A. J., Hartel, P. G., and Fuhrmann, J. J. (1995). Successional changes of bacteria in the rhizosphere of nodulating and non-nodulating soybean (*Glycine max* (L.) Merr.). 15th North American Conference on Symbiotic Nitrogen Fixation, August 13–17, 1995, Raleigh, NC.

Cole, G.M. (1992). *Underground Storage Tank Installation and Management*. Lewis, Chelsea, MI.

Cunningham, S. D., and Lee, C. R. (1995). Phytoremediation: Plant-based remediation of contaminated soils and sediments. In *Bioremediation: Science and Applications* (H. D. Skipper and R. F. Turco, eds.), SSSA Spec. Publ. 43, pp. 145–156. ASA, CSSA, and SSSA, Madison, WI.

Curl, E.A., and Truelove, B. (1986). *The Rhizosphere.* Springer-Verlag, Berlin.

Kennedy, A. C. (1994). Carbon utilization and fatty acid profiles for characterization of bacteria. In *Methods of Soil Analysis. Part 2: Microbiological and Biochemical Properties* (R. W. Weaver, ed.), pp. 543–556. SSSA, Madison, WI.

Nichols, T. D., Wolf, D. C., Rogers, H. B., Beyrouty, C. A., and Reynolds, C. M. (1996). Rhizosphere microbial populations in contaminated soils. *Water, Air, and Soil Pollution* (in press).

Pritchard, P. H., and Costa, C. T. (1991). EPA's Alaska oil spill bioremediation project. *Environ. Sci. Technol.* **25,** 372–379.

Rogers, H. B., Beyrouty, C. A., Nichols, T. D., Wolf, D. C., and Reynolds, C. M. (1996). Selection of cold-tolerant plants for growth in soils contaminated with organics. *J. Soil Cont.* **5,** 171–186.

Sasser, M. (1990). MIDI Technical Note 101: Identification of bacteria by gas chromatography of cellular fatty acids. Microbial ID, Newark, DE.

Sims, R. C., and Overcash, M. R. (1983). Fate of polynuclear aromatic compounds (PNAs) in soil–plant systems. *Residue Rev.* **88,** 1–68.

Wollum, A.G. II. (1994). Soil sampling for microbial analysis. In *Methods of Soil Analysis. Part 2: Microbiological and Biochemical Properties.* (R. W. Weaver, ed.), pp. 1–14. SSSA, Madison, WI.

Appendix 6
ALTERNATE ROUTES TO
THE THESIS

Whether you write your thesis in the traditional format or with publishable journal manuscripts, your ultimate goals are probably a degree, a publication, and a job or additional education. The thesis itself is only a part of the work that you must incorporate into your plans for the graduate program. Any one of a variety of plans can lead to your success with the thesis, and the road map in Fig. A6-1 simply represents two examples. For simplicity, the routes in that figure are assuming a master's thesis that would take approximately 2 to 2.5 years to complete and would result in the publication of one journal manuscript. A more complex doctoral dissertation or additional publications would extend the time and add to the work load but follow similar paths. Certainly the order of events may be rearranged, especially toward the last when the thesis, the defense, the job, and the publication may all close in on you at once. Be careful about entering a new degree program or taking a job before your thesis is complete.

You may feel that you are jumping through a series of hurdles as you move along your road to the thesis. For convenience in discussing the steps along the way, I'm dividing the process arbitrarily into seven hurdles, represented by the seven numbered steps in Fig. A6-1. Activities on each section of the route will overlap with each other and with those in the previous or next section. For example, you may be searching the literature, writing the proposal, and beginning the research all during the same period, but it is helpful to have some of the literature search and methods of the proposal done before you begin the research.

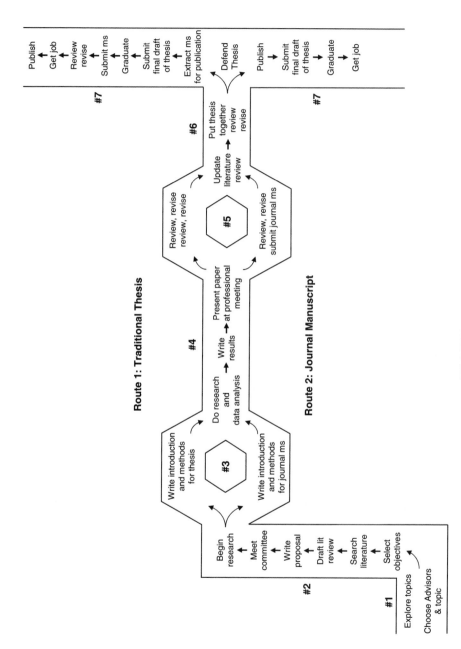

FIGURE A6-1

Alternate routes on the way to a thesis and graduation.

HURDLE 1

For your satisfaction and success, the first steps along the road must produce an inviting but practical plan. You should choose a topic for research and an advisor that you can believe in and work with intensively for at least 2 years. Before you begin this trek, you will probably investigate the graduate school you want to attend, study which areas of research are prominent in the department where you wish to enroll, and find out as much as you can about major professors working in your area. Read some of the publications by prospective advisors, and then communicate with them, go through the admissions process, and select or accept a specific major advisor.

By conducting library searches, you should explore topics that are related to the work of the professor you will work with, and when you meet with him or her, have some details in mind about the research project you want to pursue. With your advisor's help, you will consider possible hypotheses and select specific objectives for your research. At this point, having read some literature on the subject will be a real help. On the basis of what your subject involves and with the help of your advisor, choose graduate committee members and visit with each of them about your plans. Now you are ready to go to work.

HURDLE 2

With your research objectives in mind, continue to search the literature and write a draft of your literature review. This review will be updated toward the end of your program and will probably be used as a chapter of your thesis. With that rough draft behind you, write a draft of your proposal, review it, revise it, give it to your major professor for review, and revise again. Your advisor may ask that you do some preliminary experimentation before your proposal is ready for committee. When he or she approves, submit the proposal to your graduate committee and meet with them to discuss your program. Six months or more may have passed by now. (It will be a year or more if you don't move quickly, but meticulously, over these first two hurdles.)

HURDLE 3

Except for classes, research will take most of your waking hours for the next year. But don't make your last year harder or extend the time for your program by failing to write along the way. **As you begin the research, you should choose which thesis path you will take.** See Chapter 6 for further information on thesis formats. The traditional route means your thesis will ultimately consist of an introduction and chapters on literature review, methods, results, and discussion. You may have other addenda such as appendices and an abstract. This traditional thesis, as written, is seldom ready to publish when it is completed, but representative ideas and data can be extracted for publication later. The journal manuscript format means you write with publication in mind from the first, and a literature review chapter and other addenda will hold everything except the representative publishable material. To jump this third hurdle then, you need to write an introduction and a methods chapter for the traditional thesis or an introduction and methods section for a journal manuscript.

HURDLE 4

The research and the analysis of data continue, and as they near completion, you write a draft of the results. You are probably a year or two into your program. With results written, you are ready to go to a professional meeting and make a poster or slide presentation. It's time to think about prospective employers and to start making your reputation public. Continue to review and revise the various sections of the thesis. If you are writing a publishable journal manuscript as a part of your thesis, it may by now be reviewed in-house and ready to submit to the journal.

HURDLE 5

Now get serious about reviewing, revising, and finishing the thesis. For either format, supply the appendices, the abstract, and any other addenda you need. Update the literature review that your wrote more than a year ago. Review and revise each section repeatedly, without and with advice from your major professor. Review and revise again. It is not unusual to review and revise your thesis six or more times.

HURDLE 6

Overlapping of activities from step to step continue, but at some point the paths for the two thesis plans part again. You are over hurdle 6 when you are ready to defend your thesis. If you are on the journal manuscript route, your manuscript should be at least in in-house or journal review or perhaps back to you and ready for subsequent revisions and publication. For this track, be sure that complementary chapters or appendices of your thesis include all the data from your project that are not appropriate for your publication. If you have written a traditional thesis, you should by now extract representative data, write a journal manuscript, and ask for in-house reviews. Getting a draft of the manuscript ready at this point will make life less miserable as you are seeking employment and beginning a new job. Although the road behind you may have been smoother or simpler than that for your colleague who was writing the journal manuscript all along, once you submit the final draft of the thesis to committee, hurdle 7 may be somewhat harder to scale with the publication still to be submitted, reviewed, and revised.

HURDLE 7

Reviewing, revising, publishing, defending the thesis, taking exams, presenting a departmental seminar, and applying and interviewing for a job—all may confront you at once. But over this final hurdle lie your three ultimate goals: your degree, a publication, and a job or a new degree program. The order in which the tasks are accomplished can depend on your priorities and on which route you took to this point. The publication may be behind you if you took the journal manuscript route. As you approach graduation, establish priorities carefully. The job you want may become available before you finish the thesis. Don't sacrifice the degree prematurely for new short-term goals. The completed degree and a publication can be important stepping stones to your long-range plans. Remember that it is difficult to finish the thesis and produce a publication while you are becoming oriented to a new job. If you are going on for another degree, you may find that completing the thesis and a publication can interfere with scaling hurdle 1 again. Plan your strategy carefully and make a smooth jump over hurdle 7 so that reaching your final goals is a satisfying benchmark in your career.

Appendix 7
SAMPLE REVIEW OF MANUSCRIPT SUBMITTED FOR PUBLICATION

The following is a review by Dr. Gail Vander Stoep of Michigan State University that was done for a journal editor on a manuscript submitted for publication (used with permission from Dr. Vander Stoep). It has been abridged for inclusion in this appendix. The marginal notes are mine. Note how thorough and organized the review is, with comments on both general areas of content and organization as well as specific points. The reviewer accompanied these remarks with a copy of the manuscript marked clearly with even more specific details that needed to be addressed in the revision. Throughout, there is a tone of concern for the quality of the paper and a recognition of its value. Although this version of the paper was weak, note the positive comments made and the reviewer's ability to criticize and evaluate without being disparaging. The criticism is compelling but is always accompanied by suggestions for amending weaknesses. The authors of the paper expressed sincere appreciation for this review, and by incorporating the suggestions, they produced a good paper, which was subsequently revised twice and then published.

University-Based Education and Training Programs in Ecotourism
or Nature-Based Tourism in the USA
for
Journal of Natural Resources and Life Sciences Education

GENERAL REVIEW, RELEVANCE, AND RECOMMENDATION

Reviewer begins with positive comments.

It appears that the content of the article fits within the scope of the journal content. Issues of resource preservation, conservation, resource use and management, and tourism are not new, but consideration of all of these issues certainly is gaining wider attention across many sectors. University-based education that facilitates systems-based resource analysis, planning and management is critical. I would support publication of this article. However, I recommend revisions to clarify, focus and tighten the paper, as well as to make it more relevant to JNRLSE readers.

Conscious of the audience of the journal.

Improving relevance for readers

General opinions on content.

—Somewhere in the introductory and/or historical perspective sections, indicate (briefly) how ecotourism/nature-based tourism may have implications for various "natural resource" segments (e.g., natural and archaeological resources, links to agriculture-based tourism).
—Expand the "conclusions and recommendations" section, including discussion of relevance to readers, future curriculum development or revision, and the possible need for continuing education.

Title

Specific suggestion with support for her opinion.

Consider removing "and Training Programs" from the title because it appears that the scope of this study is education-based and describes individual courses..."

General Editorial Comments

Reviewer is flexible and concedes that she may not have all the best answers.

I am returning my copy of the paper with rather extensive comments written directly on the manuscript. Many of the notations are strictly editorial and deal with punctuation, capitalization, syntax, grammar, simplification of sentences, or authors may develop something even better or more in line with their thoughts.

Introduction

Another positive statement.

I like the introduction as a way to introduce the concept of and need for incorporating ecotourism concepts into university curricula. The introduction also needs to clearly define the need or rationale for the study and its results. WHY do we need to know about ecotourism/nature-based tourism curricula? Clarify why it's a problem that no data are available. (As stated, it's simply "unfortunate," as viewed by the authors.)

Historical Perspective

I believe the historical perspective section is pertinent to the paper, particularly in defining the terms ecotourism and nature-based tourism. However, the discussion could be expanded to address issues of...[*Reviewer goes into detail here on issues that should be addressed*].

Reviewer clearly rejects this discussion but has positive suggestions for its use in another publication.

I would delete the entire discussion (and Table 1) about ROS and a potential parallel with a TOS. The ideas and linkages are not fully developed; there are some "challengeable" holes in the descriptions... [*Reviewer gives specific examples*]. There is not enough space in this article to fully and logically develop the concept. (Perhaps the idea could be more fully developed for another paper for a different outlet.)

Makes a point on unity and organization.

Page 4, starting with line 19: This section seems to be more appropriate as part of the "methods" section than "historical perspective."

Research Methods

Reviewer presents ideas in the form of questions and leaves to the authors the responsibility for determining whether and how to answer them.

Perhaps a little more development of the methods section would be helpful. Some questions I had:
—Any overlap in listing of specific programs/departments among the three lists?
—What types of questions were included in the one-page questionnaire?
—How was content analysis done?
[*The reviewer continues with a dozen other questions.*]

Results

Points out the need to clarify the meaning of results.

Before beginning discussion of actual results, clearly present the response rate versus those responding who OFFER or PLAN TO OFFER a course with ecotourism content. Page 5, lines 16–19 are confusing. Comment (Page 5, lines 21–22) seems to be more of a conclusion than result.

And the need for details.

Throughout the "results," clarify what the percentages are based upon; e.g., total number of returned questionnaires, total number of schools reporting current and/or future offering of courses, total number of schools currently offering...

Reviewer points to specific parts of the paper and, without revising, helps to direct the logical thinking needed to revise.

A couple of logic concerns:
—Page 8, lines 16–18 (see comment on MS); need transition or link to the list of "other common objectives."
—Page 9, lines 8–11, and Figure 5: There seem to be different ways of interpreting "the primary focus"...actually as "the primary focus" versus simply "a component of." This is confusing.

Your use of the word "classes" is confusing. Does "class" refer to the four categories or *classifications* of approaches to instruction or to *courses?* It may be that the word is used interchangeably.

The descriptions of the four general approaches to instruction are helpful and interesting (and necessary). However, this section could be strengthened by describing HOW the four approaches were determined (perhaps this is more appropriate for the methods section?) and by indicating the number or percentages of courses in this study that use each of the approaches.

Conclusions and Recommendations

This section is the weakest part of the paper. Most of the elements currently included in the section would seem more appropriate elsewhere:
—limitation of scope and diversity: methods section. (However, a recommendation for future research and inventory is appropriate here.)
—Comments about the Ecotourism Society seem more appropriate as a sidebar or footnote.
Beginning on line 3, Page 13: It is not clear how results of the study indicate a need to improve information exchange. This may be a recommendation of the authors but is not directly linked to the results as presented.
This section could be strengthened by expanding the conclusions, adding discussion of implications for ecotourism education/ training/industry, and by providing specific recommendations. Explain WHY information exchange is important.

Figures and Tables

I suspect that some of the figures may not reproduce well, especially if reduced to one-column size. (I defer to the editors for this.)
Style guidelines for format (including capitalization) should be checked, then used consistently.

Discrepancies exist between results in Figure 2 and the "Year" column of Table 2. (Is Figure 2 needed?)
Figure 4: What were criteria for….
Other comments and questions are written directly on the manuscript.
Double check consistency of information and assumptions between tables/figures and narrative within the manuscript.

I recommend that the manuscript be revised with particular attention to clarifying the manuscript and developing a strong "implications" or "relevance" component. It could then be appropriate for publication.

Appendix 8
EVOLUTION OF A TITLE

Suppose that the following six titles are meant to describe the same study. Some are better than others, and none may be fitting for a specific manuscript. Your choice will depend on which words are most important and which best describe the study involved. Think in terms of the words and the arrangement of words that will lead a reader to the central points in your study. The publisher may also have certain criteria for titles such as length and the use of scientific names.

Sample Titles:

1. **Controlling the Bollworm**
2. **Investigations into the Effects of Several Selected Phenolic Acid Compounds on the Mortality Rate, Developmental Time, and Pupal Weight Gain of the Cotton (*Gossipium hirsutum* L.) Bollworm (*Helicoverpa zea* Boddie) in Studies Involving Larvae Fed a Synthetic Diet in the Laboratory**
3. **The Effects of Selected Phenolic Compounds on the Mortality, Developmental Time, and Pupal Weight of *Helicoverpa zea* Boddie: Synthetic Diet Studies**
4. **Benzoic and Cinnamic Acids in Synthetic Diets Retard Development of *Helicoverpa zea* Larvae**
5. **Influence of Benzoic and Cinnamic Acids on Mortality or Growth of Bollworm Larvae**
6. ***Helicoverpa zea* Larvae Response to Benzoic and Cinnamic Acids**

Number 1 might serve as a headline for an article in a newsletter for cotton producers, but it contains too little information to describe a scientific study.

Number 2 is too long, and the inclusion of all these words cannot be justified. Certainly the first three and the last three words can be omitted. Then why not "selected" rather than "several selected"? Why not "phenolic acids" rather than "phenolic acid compounds"? Why not "mortality" rather than "mortality rate"? Why not "pupal weight" rather than "pupal weight gain"? And can't we simply say "bollworm larvae" instead of "bollworm in studies involving larvae"? There is probably no need for the scientific name for cotton because we are naming the cotton bollworm and not the cotton plant. Whether the scientific name *Helicoverpa zea* Boddie appears in a title will depend upon the style of the publisher. The authority Boddie might be omitted. Or the scientific names might be used and the common name cotton bollworm might be omitted. These choices would be based upon the style of the publisher and the importance of the words for the audience.

Number 3 is also rather long. We might omit "The effects of" and "the" before "mortality," but we still have a long title and must make some other choices. Can we save "synthetic diet studies" for the abstract? Some publishers avoid two-part titles with the colon. Can we say "development" rather than "developmental time"? Or can we combine the words "developmental time and pupal weight" into a simple term such as "growth"? Again, answers to these questions depend on what we need to best describe the study and which key words will allow the readers to retrieve a publication relevant to their interests.

Number 4 is perhaps short enough but could still be improved with the omission of "in synthetic diets," unless that information is vital to a brief description of the study. This title adds a detail by naming the specific phenolics used. This information may be worth the extra space needed. However, the title breaks a convention in scientific writing by using an active verb, "retards," that describes the outcome of the study. Characteristically, popular press uses such verbs in headlines, but the scientific report simply describes results of a research effort and discusses outcomes but allows the reader to decide on any final conclusion that can be drawn from the work. What happens under the controlled conditions of a given experiment may not constitute a universal truth, and the active verb appears to be proclaiming such a truth. Some scientific periodicals use active verbs in titles, but check your own journal before you submit a title with an active verb.

Number 5 is about the same length and is similar to number 4 except it avoids the active verb and uses the common name of the species rather than the scientific name. The choice of name would depend upon which one you and your publisher believe will best communicate the information with your audience. The use of "mortality or growth" is somewhat more specific than is "development" in number 4 and may be worthy of the extra two words, especially if we can delete "in synthetic diets."

Number 6 is less descriptive of the paper's content but conserves words. Here I reserve the mortality and growth for the abstract and generalize with the word *Response*. Such a title may be the best choice, especially for display on a poster or slide.

Appendix 9
EVOLUTION OF AN ABSTRACT

The following abstract is based on studies done in the laboratory of Dr. Justin R. Morris at the University of Arkansas and is used with his permission. An original abstract from that research has been altered and lengthened with fictitious ideas and data and then pared to acceptable lengths. The examples below show four versions of the same information. Notice changes in organization, content, and wording between the drafts. Little content is lost with a reduction from approximately 370 words to 280 words and then to approximately 215 words in a third version. When I submit this third version with the manuscript to a journal, the editor may inform me that the abstract must be cut to no more than 150 words. To satisfy this request, I have to get rid of the full justification and conclusion and reduce methods and results to a bare minimum. The fourth version with approximately 140 words has lost some content but not the objectives, the most important methods, and the results. The third or fourth version might be the one published, depending on the editor's preference

Working Abstract 1 (373 words)
Evaluation of Winegrapes for Suitability in Juice Production

Indirect statement of objective.

Materials.

Abstract. A series of chemical and sensory analyses was designed to determine which, if any, of 10 winegrapes grown in 1994 in Arkansas were suitable for nonalcoholic grape juice production. Five of the 10 were classified as red grapes: Chancellor, Cabernet Sauvignon, Villard Noir, Cynthiana, and Noble. Five were classified as white grapes: Aurore, Cayuga, Chardonnay, Vidal, and Verdelet. The traditional juice grapes Niagara and Concord were used as controls for comparisons in the study.

Rationale or justification for the study.

Methods.

Note the wordiness and repetition throughout, as in "stored at 37°C" and "storage at 37°C."

Results.

Also wordy.

No overall conclusion.

Sensory quality and consumer acceptability of grape juices depend to a great extent on the process by which the juice is produced but also on the cultivar or blend of cultivars used. With today's processing techniques, winegrape cultivars may also produce nonalcoholic juices acceptable to consumers' tastes. Four different means of juice production were used to process the grapes: immediate press, heat process (60°C), heat process (80°C), and 24-h skin contact after pressing. Processed juices were sealed in 0.8-liter bottles and stored at 37°C. The juices were evaluated 1 month after processing and again after 5 months' storage at 37°C. Chemical and sensory analyses were run. Chemical analysis showed that red grapes had more acidity than did white grapes, but white grapes, except for Cayuga, had more soluble solids. Soluble solid–to-acid ratios were highest in the red grapes Noble and Cynthiana and lowest in the white Chardonnay and Vidal. Other cultivars showed no significant difference in soluble solid–to-acid ratios. Chemical analysis showed no difference within cultivar for the treatment process, except that in six of the 10 cultivars the 24-h skin contact produced more soluble solids. Consumers' preference, as represented by a sensory panel, for flavor of juices revealed greater preference for non-heat treatments, whereas the heat treatments were more preferred in color evaluations. Flavor was considered to be the most important attribute to the consumer; panelists tended to prefer those juices that had relatively higher soluble solids–to-acid ratios. The most preferred white grape juices were the immediate press of Niagara and Aurore and the 24-h skin contact treatment of Niagara, Verdelet, and Vidal. Rank preference for flavor of red grape juices did not indicate a significant preference among cultivars, thus suggesting all red winegrape cultivars were equally suited for varietal juice production.

Working Abstract 2 (280 words)

Title Shortened.

Suitability of Winegrapes for Juice Production

Rationale shorter and moved to the beginning.

Direct statement of objective.

Materials.

Methods.

Results.

Less wordiness here than in the first version.

Abstract. Recently developed processing methods may provide nonalcoholic juices from winegrapes cultivars that are acceptable to consumers. Our objective was to test 10 winegrape cultivars and four juice processes to determine their suitability for juice production. The five red grapes (Chancellor, Cabernet Sauvignon, Villard Noir, Cynthiana, and Noble) and five white cultivars (Aurore, Cayuga, Chardonnay, Vidal, and Verdelet) were compared with traditional juice cultivars Niagara (white) and Concord (red). Juices from four processes—immediate press, heat process (60°C), heat process (80°C), and 24-h skin contact after pressing—were sealed in 0.8-liter bottles and stored at 37°C. They were evaluated at 1 month and 5 months after processing. Chemical analysis showed that red grapes had more acidity, but white grapes, except for Cayuga, had more soluble solids. Soluble solid–to-acid ratios were highest in the red grapes Noble and Cynthiana and lowest in the white Chardonnay and Vidal. Other cultivars showed no significant difference in soluble solid–to-acid ratios. Chemical analysis showed no difference within cultivar for the treatment processes, except that in six of the 10 cultivars the 24-h skin contact produced more soluble solids. A sensory panel preferred flavor of juices from

Most sentences have been shortened but content is not lost.

nonheat treatments but ranked color best in the heat treatments. Flavor, the most important attribute for consumers, was rated highest in juices that had relatively higher soluble solids–to-acid ratios. The most preferred white grape juices were the immediate press of Niagara and Aurore and the 24-h skin contact treatment of Niagara, Verdelet, and Vidal. Rank preference for flavor of red grape juices did not indicate a significant preference among cultivars. Most of these winegrapes may be as suitable as juice grapes for nonalcoholic juice production.

Overall conclusion.

Abstract, Version 3 (215 words)

Suitability of Winegrapes for Juice Production

Rationale shortened even more.

Objective.

Materials.

Methods.

Results.

Notice that wording is still more concise than in version 2, but little or no content is lost.

A more specific conclusion.

Abstract. Recent processing methods may provide acceptable nonalcoholic juices from winegrapes. To determine acceptability of the juices, we compared five red (Chancellor, Cabernet Sauvignon, Villard Noir, Cynthiana, and Noble) and five white (Aurore, Cayuga, Chardonnay, Vidal, and Verdelet) winegrapes with traditional juice cultivars Concord (red) and Niagara (white). Juices processed by immediate press, heat process (60°C), heat process (80°C), and 24-h skin contact after pressing were sealed in 0.8-liter bottles and evaluated after 1-month and 5-month storage at 37°C. Chemical analyses showed more acidity in red juices, but more soluble solids (SS) from white grapes except Cayuga. The SS/acid ratios were highest from red grapes Noble and Cynthiana and lowest for white Chardonnay and Vidal with no significant differences in other cultivars. Processes produced no differences within cultivar except the 24-h skin contact produced more SS from six cultivars. A sensory panel ranked color best in the heat treatments but preferred flavor of juices from nonheat treatments. Juices with relatively higher SS/acid ratios rated highest for flavor. Acceptable white grape juices were from the immediate press of Niagara and Aurore and the 24-hr skin contact treatment of Niagara, Verdelet, and Vidal. Panelists indicated acceptance but no flavor preference among red juices. All red winegrapes tested and Aurore, Verdelet, and Vidal white winegrapes appeared suitable for juice production.

Abstract, Version 4 (140 words)

Suitability of Winegrapes for Juice Production

No rationale.

Objectives and materials combined – specific details are gone.

Basic methods still clear.

Results – Here we lose some details contained in version 3, but the most notable results are preserved.
No conclusion.

Abstract. Five red and five white winegrape cultivars were compared with traditional juice grapes Concord and Niagara to determine acceptability of juices. Juices processed by immediate press, heat process (60°C), heat process (80°C), and 24-h skin contact were sealed in 0.8-liter bottles and evaluated after 1- and 5-month storage at 37°C. Red juices had more acidity, but white grapes, except Cayuga, produced more soluble solids (SS). The SS/acid ratios were highest from red grapes Noble and Cynthiana and lowest for white Chardonnay and Vidal. A sensory panel preferred color from heat treatments but flavor of juices from nonheat treatments. Juices with high SS/acid ratios rated highest for flavor. White grape juices were acceptable from immediate press of Niagara and Aurore and 24-h skin contact treatment of Niagara, Verdelet, and Vidal. Panelists indicated acceptance but no flavor preference among red juices.

Appendix 10
PUTTING DATA INTO TABLES AND FIGURES

TABLES

Two versions of the same information appear in the tables below. Notice how Table 1b has refined the information for more clarity and ease in reading.

TABLE 1a
Percentage Germination of *Phytolacca americana* Seeds with
Hot Water and Sulfuric Acid Treatments

| Germination | Treatments | | | | |
| | | Hot water | | Sulfuric acid | |
2001	None	90 min/80°C	12 h/60°C	15 min	30 min
Rep 1	2	19	30	76	49
Rep 2	0	23	30	54	41
Avg. 2 reps	2	21	30	65	45
2002					
Rep 1	3	42	36	80	42
Rep 2	5	28	32	62	76
Avg. 2 reps	4	35	34	71	59
*Mean (both yr)	3	28c	32c	68a	52b

*Means followed by the same letter are not significantly different at the 0.05 level.

The arrangement in Table 1a does not allow us to read down from the headings to the information in the stub and field. Reps, or replications, listed in the stub are not germination as their heading states, and numbers in the field are not treatments but germination, and the reader is not told that the numbers in the field are percentages. The 2001 should appear parallel to 2002 and not as part of the main stub head above the line. Notice that the treatment times appear as mixed units of minutes and hours. In addition to these points that cloud the clarity in presentation, too many data points may distract from the point made in the table. Representative data would be better than all of them. The analysis has not been done between replications but on combined replications and years. Comparison of the means is likely all that is needed.

TABLE 1b
Percentage Germination of *Phytolacca americana* Seeds
Treated with Hot Water and Sulfuric Acid

Treatment (time/temp.)	Germination (%)		
	2001	2002	Mean*
Control	2	4	3d
Hot Water (h/C)			
1.5/80	21	35	28c
12.0/60	30	34	32c
Sulfuric Acid (h)			
0.25	65	71	68a
0.50	45	59	52b

*Means followed by the same letter are not significantly different at 0.05 level.

Table 1b improves the clarity of information by making the table read down, combining the replications as was done in the statistical analysis, using uniform units for measuring time (h), and providing meaning to the numbers in the field with the percentage sign (%). Depending on the importance of this information to the study and space available in the paper, we might use Table 1b in a publication, or we might simply provide the mean percentages and a comment on significant differences in the text without using a table at all.

FIGURES

The following two figures depict the same fictitious data. In Fig. 1a three of the lines, those for the poultry manures, are not statistically different. The symbols for all six lines in Fig. 1a are black circles, squares, and triangles, and their use

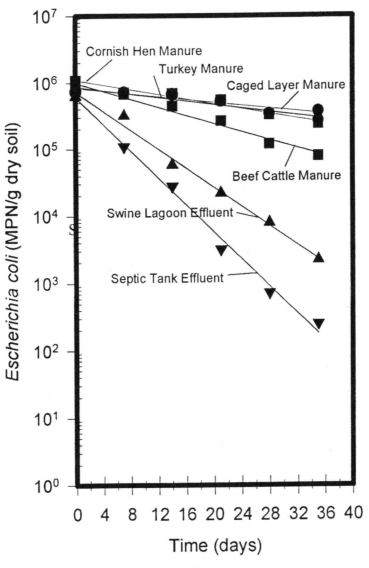

FIGURE 1a

Escherichia coli die-off in a Captina silt loam amended with cornish hen manure, turkey manure, caged layer manure, beef cattle manure, swine lagoon effluent, or septic tank effluent.

with the poultry manure data does not help much in distinguishing one line from another. The lines themselves should be the most prominent images, but the line around the box is heavier and distracting. The entire figure is compressed vertically so that the intervals on the X-axis are unconventionally shorter than those on the Y-axis. The words *Time* and *days* are somewhat redundant, and *Days* alone would be sufficient. The tic marks should not extend both within and outside the axis lines. The X-axis also contains too many numbers that run together, and they do not represent the actual days on which the data points fall. That axis should not be extended beyond the last data point. Space is also wasted in the area between 10^0 and 10^2 on the Y-axis, and because the axis is on

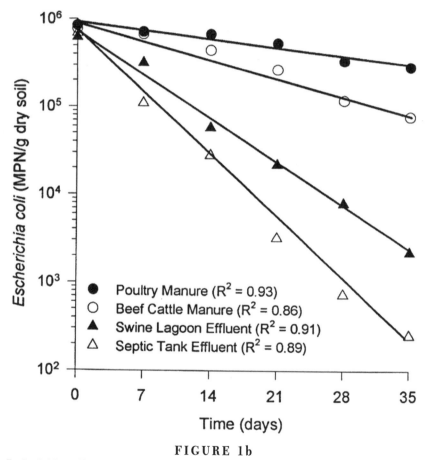

FIGURE 1b

Escherichia coli die-off in a Captina silt loam amended with poultry manure, beef cattle manure, swine lagoon effluent, or septic tank effluent.

a logarithmic scale, it need not begin at zero. Statistically, the fit of the lines with data is not apparent.

Notice the improvements in Fig. 1b. The data for all three kinds of poultry manure have been combined. In the text I will tell my audience what kinds I used and that no significant difference in die-off of *E. coli* occurred among the three. I will make clear that the R^2 represents the fit of the regression line of the geometric means for the three poultry manures. Revisions in the lines, symbols, and axes of Fig. 1b to achieve greater simplicity make the message conveyed by the data easier to comprehend. Increasing the length of the X-axis, removing the heavy box, and making tic marks more distinctive also add to clarity.

Appendix 11
SAMPLE LETTER REQUESTING COPYRIGHT PERMISSION

The following is a copy of a letter requesting copyright permission for use of the review in Appendix 7 by Dr. Gail Vander Stoep. Even though Dr. Vander Stoep has no registered copyright on this review, it still belongs to her, and permission should be requested. Note that the letter tells her where and how the material will be used and includes a copy of the version that will be published. It also assures her that she will receive credit and acknowledgment for the work.

The letter itself provides a simple form that can confer the official permission once it is signed. If my request were for a number of sections in a work or required lengthy description, I would devise a form to list information on materials requested. Both the *CBE Style Manual* (5th ed., 1983) and the *ACS Style Guide* (Dodd, 1997) provide samples of forms that may be used. See Chapter 12 for complete references to those publications. For the author's files, a duplicate of the request with an original signature by the person making the request is enclosed.

Department of Agronomy
University of Arkansas
Fayetteville, AR 72701
13 May 1996

Dr. Gail Vander Stoep
Michigan State University, PRR Department
131 Natural Resources Building
East Lansing, MI 48824-122

Dear Dr. Vander Stoep:

I am preparing a handbook, entitled *Scientific Papers and Presentations*, to be published by Academic Press, Inc., San Diego, California, in 1996. The book will be used by graduate students in the sciences and by career scientists. To illustrate points that I make in the text, I plan to use several appendices. I want one of these to be the review you did for the *Journal of Natural Resources and Life Sciences Education* on the manuscript entitled "University-Based Education and Training Programs in Ecotourism or Nature-Based Tourism in the USA."

Can you please grant me permission to reproduce an abridged version of your review in this book and in any subsequent editions or reprints thereof. I enclose a copy of the version I want to publish.

With my use of the review, full credit and acknowledgment will be accorded to you. Please confirm your permission for me to use this material by completing the form below and returning this sheet to me. I enclose a duplicate for your files.

Sincerely,

Martha Davis

Permission is granted for use of the materials described in the letter above.

Date_____   Signature_____

Title_____

Appendix 12
USE OF COLOR IN
VISUAL AIDS

The following excerpt is taken from Chapter 4, "The Theory of Colors" in *Cartographic Relief Presentation* by Eduard Imhof. (Permission to use this material has been granted by Walter de Gruyter, Hawthorne, NY.) I would certainly recommend that anyone interested in the use of color, consult Imhof's book. Imhof's expertise is in the area of cartography, but what he says about color can apply equally to its use in any visual presentation. In reading the following, you may substitute the term **visual aid** for **map** and think of these principles and rules in regard to making slides, a poster, or other colored displays. Notice the emphasis on the following points:

—**Subdued colors, neutral colors, or those toned down with grey are best for backgrounds and large areas.**
—**Bright colors and contrasting colors are best reserved for highlighting points of emphasis or for small areas.**
—**Combinations and juxtaposition of colors require that the choices be complementary and harmonious.**
—**Individual choices allow the creator of the map (visual aid) to articulate his or her point of emphasis and make the communication effort unique.**

IMHOF'S THEORY OF COLORS
On the Harmony of Colors and their Compositions

The eagle has a keener eye than man, but all that he derives from his perceived image is whether or not it is of interest to him, perhaps edible or dangerous.

The human being, although, as a rule, more highly developed in an intellectual–spiritual sense, views the visually perceived world—the forms and colors of things—primarily as a psychological experience. He hangs a colored picture of red roses above his bed. But when the same colors appear in the picture of an open wound, he is horrified.

A color in itself is neither beautiful nor ugly. It exists only in connection with the object or sense to which it belongs and only in interplay with its environment. Concepts of harmony, accord and melody always refer to composition—that is, to the way in which they harmonize or interplay with acoustics or visual elements. With a finely developed artistic sense or as a result of careful education, we can, of course, learn to perceive even non-objective, abstract color compositions as beautiful or ugly. Many simple heraldic figures are, for example, included in such abstract compositions.

Attempts have often been made, and are still being enthusiastically made today, to evaluate drawing and painting by scientific methods. This, however, is very difficult. Here one cannot expect to find laws which can be proven but, rather, demonstrations of generally valid perceptions, experiences, and fashions.

It has been shown repeatedly that great artists have broken through the accepted laws of composition and, in so doing, have achieved extraordinary effects. In the field of painting, intuitive, artistic perception takes over where scientific logic fails.

In spite of these reservations, there follows here an attempt to set down several general *rules of color composition,* insofar as they are significant for maps.

First to be examined are the *combinations of two or more colors,* taken outside the context of pictures or compositions. The latter will follow later.

a) **Combinations of two or more colors.** Which colors, for example, if presented as adjacent, but separate, similar rectangles or rectangle-like areas, would be most pleasing, and which would clash or be out of harmony? The answers to such questions, vary greatly as a rule from one case to another. Fashion, education, psychological inclination and the emotional state at the time are as important here as the existence of any artistic inclination.

In general, the unbroken sequence of colors from the color circle is perceived as harmonious. Compounds of only two colors have harmonious effects if they are complementary colors, that is, if they lie opposite each other in the color circle. The same principle holds for groups of three. Examples are:

Groups of two: Yellow–violet
                          Yellow orange–blue violet
                          Orange–blue
                          Red orange–blue green
                          Red–green
                          Red violet–yellow green

Groups of three: Yellow–red–blue
Yellow orange–red violet–blue green
Orange–violet–green
Red orange–blue violet–yellow green

Such duos and triads are even more harmonious if their colors are lightened by white, darkened by black or toned down to a pastel shade by grey, to equal extents. *Subdued colors are more pleasing than pure colors.*

Brown colors are composed of yellow and red, and a small amount of blue. Their harmonious complements are found in colors in which that tone is dominant which is weakest in the brown in question. Examples:

Yellowish brown–blue violet
Reddish brown–blue green

In general, experience has shown that harmony exists between two colors when their subtractive mixtures produce black or grey and, correspondingly, their additive mixtures produce white or grey. This holds true for every pair of complementary colors and, hence, for any two colors whose numerical values in the Hickethier color classification yield, together, a sum consisting of three equal digits.

Leonardo da Vinci said, "There is no effect in nature without cause." Therefore, one searches for explanations of the color harmonies described above. One could find them perhaps in an unconscious striving for order, in the complementary character of colors, or in their contrasts. Perhaps, however, the causes of harmonious effects lie deeper, perhaps in our familiarity with the colors of daylight. We perceive a color composition to be well integrated or harmonious when it results in the white of daylight if mixed additively, or produces grey when mixed subtractively. This applies for the groups listed above. As we have already established, two adjacent colors blend mutually as far as our sense of sight is concerned, and each color tends toward the complementary color of the neighboring hue. *Psychological color perception always tends, therefore, in the direction of composing complementary colors.* This statement appears to be quite significant in explaining the perception of harmony discussed here.

For the same reasons, perhaps, two or more different, bright colors — placed in close proximity to each other — have an unharmonious effect when their combination does not make up white, grey or black.

Examples of this type are the following groups of two:

Red and violet
Violet and blue
Blue and green
Strong yellow and pale whitish red

One of the most troublesome area colors, for cartographic and other purposes, is *yellow*. It can provide good effects, when it is used in carefully calculated amounts, to give a "warming" effect to *whole* areas as a base of background tone, but is poor in contrast with white or pale, desaturated bright, yellow-free areas. Yellow and white are similar in that they are the light components of various illumination sources. They stand out poorly when placed adjacent to one another. Midday light and twilight do not appear in a landscape simultaneously. On the other hand, however, yellow goes well with blue, violet, and blue-brown probably because of the effects of contrast. Yellow light produces blue or violet shadows.

White, grey and *black* — as *"neutral"* colors, as a surfeit of or lack of daylight — go well with all the bright colors. The compatibility of white and yellow, described above, is the only exception. It is emphasized here once again that the clashing effects of other colors can be subdued by grey, black or dull brown intermediate tones. Unfortunately, however, in maps we can seldom make use of this facility.

So far we have dealt only with the colors themselves and have not gone into *area/size relationships in connection with colors* and into the various color intensities. The harmony of the colors can be considerably reduced, or even improved, if their *areas* are *unequal in size* and if the colors are of unequal intensity. *Relationships exist between color intensities and area dimensions.* The purer and richer a color, the smaller its area should be. The duller, the paler, the greyer, the more neutral the color, the larger the area which can be covered. Two bright colors in areas of unequal size go well together only when the smaller area component is strongly colored and the large area is weakly colored. Not only are the qualities of the individual components important, but their quantities as well.

Of a completely different type and based on other phenomena and conventions is a second group of color combinations with harmonious effects. It consists of the *sequence or the change of several continuously graduated colors of one and the same hue* [sic] gradation sequences such as these are brought about by the successive addition of white, grey or black — or even of another bright color.... These admixtures give desaturated series, pastel series, shaded series. In nature, they frequently result from differences in distances of observation, or in light intensities, or through atmospheric haze. We perceive them as pleasing or harmonious, perhaps because they have an ordering, grouping, connecting, calming effect and to a large extent reflect the environmental appearances to which we are accustomed.

b) Color compositions. "Tones, harmonies, chords, are not yet music" (Windisch). Only the composition as a whole determines the good and bad of a piece of graphic work. This is also true of a map. Here, of course, one is not completely free to create graphic form. Nevertheless, the cartographer should not blame the chains that bind him for any lack of taste in his work, because he also has sufficient alternatives available to allow his good aesthetic judgment to be employed.

There follow several empirical *rules* which are especially applicable to map design.

First rule: Pure, bright or very strong colors have loud, unbearable effects when they stand unrelieved over large areas adjacent to each other, but extraordinary effects can be achieved when they are used sparingly on or between dull background tones. "Noise is not music. Only a piano allows a crescendo and then a forte, and only on a quiet background can a colorful theme be constructed" (Windisch).

The organization of the earth's surface facilitates graphic solutions of this type in maps. Extremes of any type—highest land zones and deepest sea troughs, temperature maxima and minima, etc.—generally enclose small areas only. If one limits strong, heavy rich and solid colors to the small areas of extremes, then expressive and beautiful colored area patterns occur. If one gives all, especially large areas, glaring, rich colors, the pictures have brilliant, disordered, confusing and unpleasant effects.

Second rule: The placing of light, bright colors mixed with white next to each other usually produces unpleasant results, especially if the colors are used for large areas.

Third rule: Large area background or base-colors do their work most quietly, allowing the smaller, bright areas to stand out most vividly, if the former are muted, greyish or neutral. For this very good reason, *grey* is regarded in painting to be one of the prettiest, most important and most versatile of colors. Strongly muted colors, mixed with grey, provide the best background for the colored theme. This philosophy applies equally to map design.

Fourth rule: If a picture is composed of two or more large, enclosed areas in different colors, then the picture falls apart. Unity will be maintained, however, if the colors of one area are repeatedly intermingled in the other, if the colors are interwoven carpet-fashion throughout the other. The colors of the main theme should be scattered like islands in the background color (see Windisch, 318).

The complex nature of the earth's surface leads to enclosed colored areas, like these, all over maps. They are the islands in the sea, the lakes on continents, they are lowlands, highlands, etc., which often also appear in thematic maps, and provide the desirable amount of disaggregation, interpretation and reiteration within the image.

In this respect, great importance is laid on delineation of areas within maps on the selection of sections, and even on the combination of maps in atlases and also on map legends. Cleverly arranged, legends put life into empty spaces, loosen up uninteresting parts of the image and often produce a balance in the composition.

Fifth rule: The composition should maintain a uniform, basic color mood. The colors of the landscape are unified or harmonized by sunlight.

In many maps and atlases a special green printing color is used for low lying land apart from the blue of the seas. The impression produced is usually poor, the colors of the oceans separating sharply from those of the land areas. In the "Schweizerischer Mittelschulatlas" ("Swiss Secondary School Atlas") and in other maps, however, the low-land green is produced, by over-printing, from the light yellow (used to cover all land areas) and the blue of the oceans. Other mixed tones are also derived from a few basic colors. Only in this way can the unity of light and tone be introduced and disturbing color contrasts avoided.

The idea of a single, uniform basic color mood should not be exaggerated, however. The freshness of colors should not suffer and the contrast between them should not be unduly subdued. The whole map sheet should not appear dull, jaundiced and dead.

Sixth rule: Closely akin to the requirement for a basic color mood, mentioned above, is that of a steady or gradual de-emphasis in colors. A continual softening of area tones is of primary importance in cartographic terrain representation. The natural continuity of the earth's surface demands a similar continuity in its image. Aerial perspective gradation helps this to be attained.

The principle is in no way opposed to a contrary requirement, that of contrast effects. A master reveals himself through the way in which he manipulates the different principles, using moderation on the one hand, but applying deliberate and carefully considered emphasis on the other.

In all questions of form and color composition, one should strive for simple, clear, bold and well-articulated expression. The important or extraordinary should be emphasized, the general and unimportant should be introduced lightly. Uninterrupted, noisy clamor impresses no one. Activity set against a background of subdued calm strength produces a deeply expressive melody.

This is also true in maps. The map is a graphic creation. Even when it is so highly conditioned by scientific purpose, it cannot escape graphic laws. In other fields, art and science may take different pathways. In the realms of cartography, however, they go hand-in-hand. A map will only be evaluated as good in the scientific and didactic sense when it sets forth simply and clearly what its maker wishes to express. A clear map is beautiful as a rule, an unclear map is ugly. Clarity and beauty are closely related concepts.

Appendix 13
SAMPLE SLIDES AND SLIDE SET

Probably the most common problem with designing slides is overloading them with too many words or images. The following examples may help you to recognize how to limit the amount of material that the audience must view and digest as you are talking. In the first two examples, notice that the same basic information is presented. In Example 1, the words "Characteristics of" are superfluous and complete sentences are unnecessary. In Example 2 only the key words are presented. If the speaker assumes the responsibility for presenting the full sentences orally, the audience can better comprehend the key words and listen to the speaker at the same time. Example 1 places a greater burden on the viewers in that they must read through the slide and select those points. Simplify each of your slides as much as possible.

Example 1

> ### Characteristics of Spinach
> *(Spinacia oleracae)*
>
> \>Spinach originated in Central Asia.
> \>It is valued at over 200 million dollars.
> \>It is dependent on temperature and photoperiod.
> \>It is classified as savoy (thick, crinkly leaves), semi-savoy, or flat-leaf.

Example 2

> ### Spinach
> *(Spinacia oleracae)*
>
> \>Origin — Central Asia
> \>Value — $200 million
> \>Temperature and light dependent
> \>Types — savoy, semi-savoy, flat

The principle of slide simplicity is illustrated further in the following figures. The flow chart is a good idea, but too much information greets the audience at once if we simply present the entire figure on one slide as in slide 5. Feed the viewers the information gradually by building up to slide 5, showing particular sections as you describe the content in them, and then presenting the composite slide only after you have guided them through the information. I call this method the house-that-Jack-built technique of slide composition. Notice in slides 3 and 4 that the boxes for ammonium and nitrate are kept in the same locations as in slide 1. The location and space are communicating, and we effectively break the convention of balance in this group of slides.

That technique can be used with a series of slides, or it can be designed with animation in an electronic presentation. Animation can be useful if it is not overused or used erratically. The following slide illustrates Neil Postman's analogy of coloring a semantic environment and could effectively be presented with

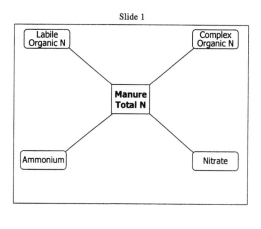

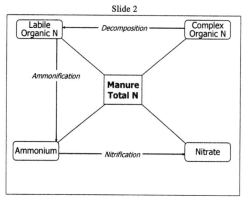

Slide 3

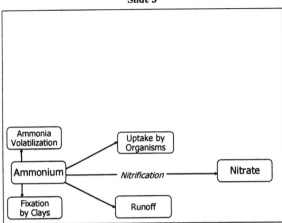

Slide 4

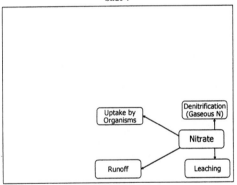

Slide 5

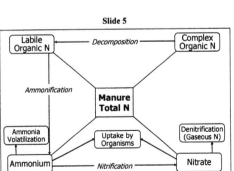

animation as you present the first, then second, and finally the third section of the slide. Excessive motion such as flying the images in, bringing them across the caption, or taking time to slowly raise the vial and rotate it to pour the color element into the flask could draw too much attention to the image itself and away from the idea it illustrates. Quite simply, all should come from the right, or the first from the left and the other two from the right. The principle of simplicity applies to animation the same as it does to other elements in communication.

The following slide set, which was used in a presentation based on the preliminary study described in Appendix 5, adheres to that principle of simplicity. Slides were all in color, and the author, Terry Gentry, chose a subdued dark blue for the background and yellow and white for the text. The photographs and figures add other colors to highlight the presentation. The word slides serve as the outline for the speaker and contain only enough information to reinforce what he will say to the audience. His repeated reference to the literature adds credibility to his own ideas about phytoremediation and microbial diversity in the rhizosphere. Notice how photographs are clearly related to the text and are mixed throughout the set to help convey the message and to provide relief from the word slides. His results are presented entirely with the table and two figures, which he can leave on the screen as he discusses the outcome of his study. These 24 slides are not too many for Terry's 12- to 15-minute presentation (Fig. A13-1). Half this number would be more appropriate if more explanation of complex data were needed.

2

RHIZOSPHERE BACTERIAL DIVERSITY RELATIVE TO PHYTOREMEDATION OF ORGANIC CONTAMINANTS

Terry J. Gentry

**Department of Agronomy
University of Arkansas**

3

PAHs
Polycyclic Aromatic Hydrocarbons

Phenanthrene   **Pyrene**

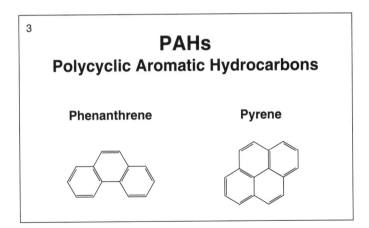

4 # Pathways of Dissipation

- Volatilization
- Irreversible sorption
- Leaching
- Accumulation by plants
- Biodegradation

Reilley et al., 1996

6 # Remediation Techniques

- Physical containment
- Excavation and treatment
- *In-situ* treatment

Lee et al., 1988

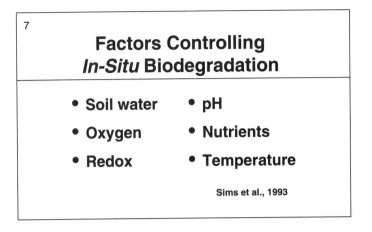

Factors Controlling
In-Situ Biodegradation

- Soil water
- Oxygen
- Redox
- pH
- Nutrients
- Temperature

Sims et al., 1993

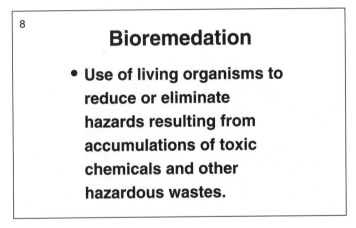

Bioremedation

- Use of living organisms to reduce or eliminate hazards resulting from accumulations of toxic chemicals and other hazardous wastes.

Phytoremedation

- Use of green plants to remove, contain, or render harmless environmental contaminants.

Cunningham & Berti, 1993

10
Phytoremedation
If pollutants are:

- Near the surface

- Relatively non-leachable

- Not imminent risk to health
  or environment

Cunningham & Lee, 1995

12
Rhizosphere

- Zone of soil under the direct
  influence of plant roots and in
  which there is an increased
  level of microbial numbers
  and activity.

Curl & Truelove, 1986

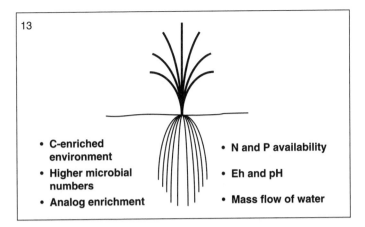

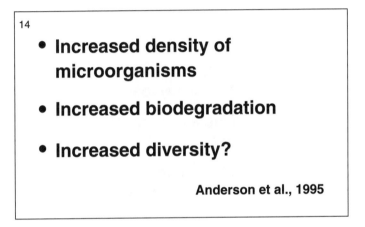

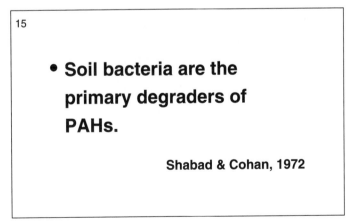

Objective

* To assess the impact of the rhizosphere on soil bacterial diversity

Materials and Methods

* Captina silt loam and Appling sandy loam

* Bahiagrass and no plant control

* Growth chamber – 3 wk 16/8 h and 27/16 ±1°C

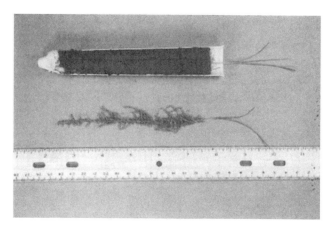

19

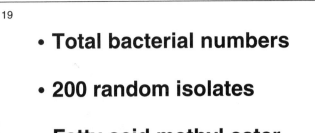

• **Total bacterial numbers**

• **200 random isolates**

• **Fatty acid methyl ester analysis (FAME)**

20

Total Bacterial Numbers

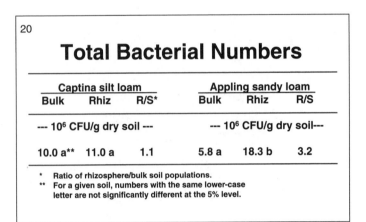

Captina silt loam			Appling sandy loam		
Bulk	Rhiz	R/S*	Bulk	Rhiz	R/S
--- 10^6 CFU/g dry soil ---			--- 10^6 CFU/g dry soil---		
10.0 a**	11.0 a	1.1	5.8 a	18.3 b	3.2

*　Ratio of rhizosphere/bulk soil populations.
**　For a given soil, numbers with the same lower-case
　　letter are not significantly different at the 5% level.

21

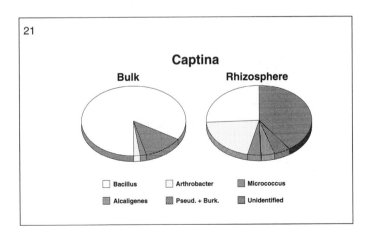

Captina

Bulk　　　　　　　Rhizosphere

☐ Bacillus　　☐ Arthrobacter　　▨ Micrococcus
▥ Alcaligenes　　▩ Pseud. + Burk.　　■ Unidentified

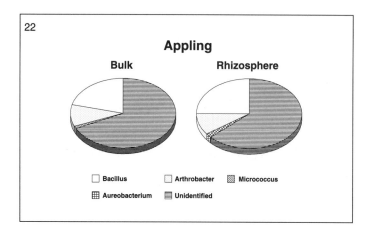

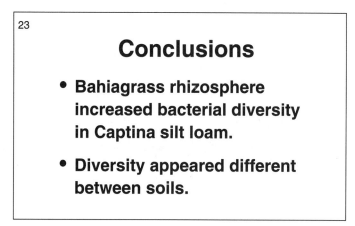

23

Conclusions

- Bahiagrass rhizosphere increased bacterial diversity in Captina silt loam.

- Diversity appeared different between soils.

24

- Increased understanding of the rhizosphere influence on bacterial populations may enhance remediation of PAH-contaminated soils.

Appendix 14
ORAL PRESENTATIONS AT MEETINGS

When you read the following editorial, you may feel that Jay Lehr is being a bit severe in suggesting punishment by stoning for inept speakers. But after you have traveled several thousand miles to attend meetings to enhance your scientific and professional development and have listened to all too many speakers who waste your time with mumbling, reading, showing unreadable slides, or running long minutes overtime (perhaps into your own presentation time), then you may agree with Lehr's pronouncement. At least read the positive suggestions he makes to achieve a successful presentation; he hits upon the main ideas about the audience, the subject matter, the visual aids, and the speaker's delivery. Perhaps his stoning is facetious; his ideas for a good presentation certainly are not.

Editorial[1]
Let There Be Stoning!
by Jay H. Lehr

Let there be an end to incredibly boring speakers! They are not sophisticated, erudite scientists speaking above our intellectual capability; they are arrogant, thoughtless individuals who insult our very presence by their lack of concern for our desire to benefit from a meeting which we chose to attend.

---

[1]Reproduced with permission from Marilyn Hoch, senior editor of *Ground Water*. This editorial by Jay H. Lehr appeared in *Ground Water* **23**, 162–165.

We attempt to achieve excellence of written presentation in our journals. We can require no less in our conferences. It is an honor to be accepted as a speaker who will spend the valuable time of hundreds of scientists at a conference. Failure to spend this time wisely and well, failure to educate, entertain, elucidate, enlighten, and most important of all, failure to maintain attention and interest should be punishable by stoning. There is no excuse for such tedium, so why not exact the ultimate penalty?

Not long ago I became so enraged by a speaker at a conference I moderated, that I publicly humiliated him before 200 hostile attendees. This young man chose to read in a monotone from a secretarial pad, flipping pages for 30 minutes of a scheduled 20-minute speech while complex slides tripped incomprehensibly across a screen behind him. At the conclusion of this group insult, he had the nerve to summarize his presentation, looking up for the first time, by stating that he hoped he had helped us to understand the relationship between the rain in Spain and the crumbling of the Rock of Gibraltar, or some such ponderous chain of reasoning. As I awakened the remaining audience, who had not the nerve to walk out as others had, I explained to the young man that he had done no such thing. Trembling as I spoke, I told him and the audience that his paper was an insult which had obviously bored and irritated a kindly group of scientists who deserved better. Those who kept awake refrained from stoning him, though they surely had adequate cause. The young man collapsed into his seat in shock as I proceeded with my vocal condemnation, the audience was pleasantly aghast, and this editorial was born.

What I said then I write now. It needs saying and writing. We have all experienced this insult, and many of us have been guilty of purveying it. It must stop. It is not funny. The penalty must be severe.

I recognized the problem when attending my first conference with my thesis advisor as a graduate student in the '50s. I was appalled at the dreadful presentation I was subjected to. The professor tried to calm my immature ravings by explaining that all meetings were like this and that their value was in the halls, not the auditoriums. I could not see why value could not be attained in both places, but I have remained quiet too long. And so to begin.

The average conference paper is 20 minutes in length. It is not a college lecture where students are to absorb the minutest detail of a subject planned and presented as part of a 10 to 16 week curriculum. Rather, a conference paper offers an up-to-date capsule summary of a particular piece of ongoing or completed research for the purpose of bringing fellow scientists up to date on activities in their field.

A speaker cannot hope to teach the audience the specifics of his work, but he can elicit a valuable appreciation of the research effort and imply the value of the contribution to the growing body of knowledge on the subject. To achieve this he must convey enlightened enthusiasm for his subject and the advances he has attained.

Without exception a presentation with the aforementioned goals can and should be made extemporaneously. A scientist who cannot retain in his head the essence of his latest work can hardly be said to be enraptured by his subject. If a speaker is not excited enough by his area of expertise to weave it comfortably into the fabric of his cognitive thought processes, then how can he hope to excite an audience to an acceptable level of appreciation?

There is never an excuse to read a paper. True, it is the rare speaker who can articulate verbally the same elegant phraseology he commits to paper with the benefit of editing, but fewer still are the preachers of science who can bring the written word to life and the audience to the edge of their seats. Better to lower the level of verbal excellence and raise the level of extemporaneous energy. The audience will never know what perfect phraseology they are missing, and the speaker must not allow himself to be frustrated by the inability to turn a perfect phrase in the air. In any case, a paper written for publication never reads well out loud. It's really a different medium. If the speaker excites the audience with his energy, they will want to read the paper later, and then they can rapture in the precision of the written word.

A few notes or an outline are all that are required to maintain order and organization. Slides, of course, can serve the same purpose, but never subject your audience to poor slides just because they serve as an outline for your talk. Poor slides are just a distraction from your hopefully vivid words. They must be brightly lit and convey a simple thought. If you need a pointer to indicate an important concept or location on a slide, it is probably too crowded or difficult to comprehend. If you can't read the print on a slide clearly with the naked eye (reading glasses are permitted) when holding it in your hand, it is inadequate for viewing with a slide projector in any size room with an adequately sized screen.

Never, but never (remember stoning) show a slide and then apologize for it. Don't show it. What did you think of the last speaker you heard say "I apologize for the poor quality of this slide," or "I realize no one past the front row can read this slide," or "I'm sorry you can't read the columns of numbers on this slide but I just wanted to point out..."? Point out what, fella—we can't read it, remember? Well, what did you think of these speakers? Dumb at best; *&!*&!*@*, at worst! Resist! Resist! Don't show bad slides! They never help; they always hurt. Don't be afraid to use no slides. Word pictures can be great if you practice painting them with a bit of rehearsal. Many of the best college professors you've heard do just fine in their lectures without slides. You can too— kick the crutch! But if you want to use slides, make them good ones. Good ones are not cheap. You can easily spend a few hundred dollars on a good set of slides for a talk, but look at the dividends:

—Your audience will sit up, take notice, and think you're great, someone special!

—You will invariably find many opportunities to use good slides over and over.

—Your audience paid good money to come and hear you. You probably got in free or at a reduced rate. Reinvest the savings in good slides and give your audience a dividend on their investment.

Don't stay on one slide too long; put blanks between slides if you have a lot to say before the next slide. The old slide is distracting. Don't let the slide lull you into a monotone; keep a high energy level with lots of enthusiasm. When you are giving a paper, you are an actor on a stage. You may be an incredible dullard in real life, putting people to sleep right and left, but at that podium, you're a star. You're an entertainer, an educator; put on a happy face and kick ass...or get off that stage. Science is sensational; working in a factory is boring; seeking scientific truth is a turn on, so turn on or you'll turn your audience off. You ought to know, your colleagues have been doing it to you for years. Dare to be different. Use your hands, move around, not to the point of distraction, but look alive! Unless you're a pro and I'm boring you with the obvious, rehearse. Rehearse before a friend or relative and to yourself in the quiet of your mind, on a drive, a run, a swim, a cycle, a daydream, anywhere! Listen to yourself. Your wife, kids, and friends won't want to listen to you; bribe them, they will. If you tell them to be tough on you and let you know what's really bad, they'll love it. Think of the time the audience is collectively giving you. One hundred people times 20 minutes is 33 hours. Don't you owe them a few hours of effort in return?

Get your timing down. No one minds your going a minute or two overtime, but five or eight is inexcusable. Face it, there is extraneous material in your talk. You may love it, but the audience can do without it. Get to the point earlier, and spend more time on the meat and less on the soup and nuts. In the beginning, tell the audience what you're going to tell them. Then tell them, and be sure to leave time at the end to tell them what you told them. It sounds simple, but it works and they will appreciate it.

Make sure you talk into the microphone; tell the audience to let you know if you're too loud or not loud enough. You will lose 20 seconds regaining your composure and properly modulating your voice, but that beats 20 minutes of deafening silence or a rumbling sound system.

Avoid jokes unless you're a stand-up comic. Nothing is colder than a failed attempt at humor. If there is anything humorous in your subject, milk it. That's real and will be well received.

When all is said and done, more is said than done. Don't waste words, but if you must, remember that attitude is 75 percent of nearly everything in life; enthusiasm is at least that in public speaking. Brim with enthusiasm; if you don't have it for your work, how can an audience have it for you? Come alive!

A few words for moderators — you're the master of ceremonies, and you can set the tone for all of the speakers. Show an interest in the session. Open with 30 seconds of well-planned comments. Introduce each speaker with five pertinent points of information which you committed to memory in the past 10 minutes, i.e., college degrees and colleges attended, two significant past work affiliations (if pertinent), current work affiliation and activity focus. Do it like you know the speaker well, even if you never laid eyes on him before. You can do it. It takes just three minutes to learn five facts for a short duration. If you're not willing to put in the time, don't accept the job of moderator. When the speaker finishes, keep order during the question period and don't hog the microphone yourself, but do tie it all up with 30 seconds of concluding remarks, if appropriate. A good moderator can really help; a bad one gets in the way, wastes time, and impedes the performance of the speakers he is there to assist.

When on the speakers' platform, unless you have a natural wit and air of showmanship, you cannot afford to be yourself. You must be an actor who is privileged to educate and entertain. The latter must come first or the former has no hope of attainment. But here is one simple rule that can make us surprisingly as comfortable before a group as with a single friend. **Be intimate with your audience.** Make them feel that you are there because you care about informing each and every one of them; no matter if there are 40 or 400, **be intimate.**

When you see the rare speaker who has an air of showmanship that allows him to get into the minds of his audiences, do you comment on how lucky the speaker is to be a natural? Are you sure you can't hope to emulate such a performance? A stage is meant to be acted on, whether to perform in a play or exhort a college student into a broader and deeper understanding of the subject at hand.

We have classrooms in college and stages at conferences because we know that the learning process can be enhanced by animated oral presentations which transcend the capacity to learn from the written page. Unfortunately, most of us achieve less, not more. We deliver an unenthusiastic reading or account which falls more deafly on ears than dead prose fall on our eyes.

Don't get up and do what comes naturally if what comes naturally is a dull, witless, monotonous presentation of unexciting facts. If your work is in fact dull and unexciting, don't burden any audience anywhere with a conference presentation. Publish somewhere if you must, if you can!

If, on the other hand, your work has substance that can be brought to life, do just that. Waste no more time saying you can't do it. Do it or have a colleague do it. There is no longer any excuse to be dull. Regardless of the fact that TV evangelists have given enthusiasm a bad name, **be enthusiastic!**

I studied astronomy under a dullard and thought it was a dead science. Carl Sagan taught me differently. I studied biology under a bore and saw no future in it in my mind. Paul Ehrlich showed me differently. My first economics professor put me to sleep, but Paul Samuelson awakened my interest.

I became a geologist because my earliest mentors, Cary Corneis, John Maxwell, and Harry Hess made the earth live for me. Make your subject — no matter how esoteric — live for your audience if only for 20 minutes.

If everyone takes my message to heart, there need be no more public humiliations and even fewer stonings. But if egotistical, pompous, cavalier, obtuse, inconsiderate ignoramuses insist on ignoring these words to the wise, let there be stoning!

Appendix 15
SAMPLE TEXT FOR POSTER

The following poster layout and text are from the poster pictured in Fig. 17-2 on **page 196.** The authors had a large (approximately $4' \times 8'$ or 1.2×2.4 m) board for display. A smaller board would require that the text be reduced in a manner similar to that of reducing an abstract (see Appendix 9). Probably with a board half this size, the introduction might be no more than one or two sentences of justification preceding the objectives. Methods and results would have to be cut to retain only essential information. Reference to the literature would probably be omitted or very limited, and discussion might be no more than a sentence or two.

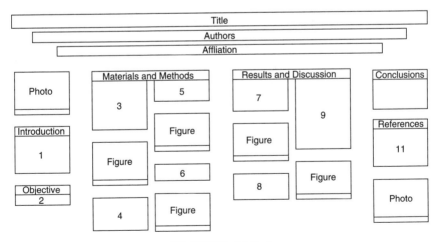

FIGURE A15-1
Layout for poster in the photograph (Fig. 17-2) in Chapter 17.

Notice the layout of this poster. Sections are balanced with the first and last columns as single rows. The center portions also balance in the same way, with double rows connected by their headings. The fact that the columns are not the same length and each piece of the poster is not the same size adds aesthetic interest that would not be present in a monotonous repetition of the same sizes and levels throughout. Similar layouts can be done for posters printed on single sheets; see the spacing, arrangement, and sizes in Fig. 17-3, Chapter 17. Note here that spacing within sections communicates; spaces are wider between sections, but both columns within the materials and methods and the results and discussion sections are grouped closer together.

The text is presented below with numbers corresponding to those on Fig. A15-1. It is in short segments, with only section 9 being somewhat long, and much of both the methods and results is communicated via figures. Clear statements of objectives and conclusions are set apart with their own headings to make it easy for the reader to know what was planned and what was the basic outcome even without reading methods and results. Photographs help to describe the site and the system, but they also have aesthetic value and serve as relief from additional text that might be needed to describe them.

Notice that the central idea, as expressed in the objectives and conclusions, is carried into detail with photographs, figures, and text to make this poster a reader-friendly display.

TEXT FOR FIGURE A15-1:
(This research was conducted in the late 1980s)

RENOVATION OF ON-SITE DOMESTIC WASTEWATER IN A POORLY DRAINED SOIL

D.C. Wolf, E.M. Rutledge, M.A. Gross, and K.E. Earlywine
University of Arkansas and Arkansas Department of Health

1. Introduction

Justification

Approximately 29% of the U.S. population disposes of domestic wastewater through onsite disposal systems (Miller, 1984). During the past 10 years, the rate of construction of soil absorption systems for onsite wastewater disposal has increased by 20% (Reed et al., 1989). However, septic systems are frequently cited as sources of water contamination.

Introduction to topics in the poster

To protect water quality, it is essential that the soil renovate the domestic wastewater before it replenishes water supplies (Reneau et al., 1989). In poorly drained soils, hydraulic limitations can result in septic systems that fail to adequately purify the effluent and thus impose an adverse impact on water quality.

2. Objective

Specific plan

The objective of the study was to use a low pressure distribution system with tile drains to evaluate the renovation of septic effluent in a poorly drained soil.

3. Materials and Methods

Methods are important and unique to this study; therefore, they are described in detail with three figures to illustrate.

A clear statement of methods provides credibility for the study.

Septic effluent was pumped from the dose tank into filter field absorption beds indicated as B1, B2, B3, and B4 in Fig. 1. The beds were loaded at a rate of 18 L/m^2/day and received the septic tank effluent through 0.48-cm orifices in 3.8-cm nominal-diameter, schedule 40 PVC pipe. The system distributed effluent evenly over the beds by maintaining approximately 60 cm of head. Tile drain trenches shown as T1, T2, T3, T4, and T5 in Fig. 1 were located beside and between the absorption beds. A cross section of the system is shown in Fig. 2. The drain trenches and the absorption beds were separated by 100 cm of undisturbed soil. Each sand-filled tile trench contained a nominal 5-cm-diameter Hancor "Turflow" slotted drain pipe located 10 cm from the bottom of the trench. The tile drains discharged into a sump where each tile was sampled. In addition to the filter field system, background tile drain trenches denoted as T6, T7, T8, T9, and T10 were sampled.

Description of the site

4. The study site had a 0% slope and was located on a Calloway silt loam (fine-silty, mixed, thermic Glossaquic Fragiudalf). Poor drainage and a high seasonal water table were indicated by mottling throughout the soil profile, which had a fragipan or Btx horizon at 116 cm. The amounts of precipitation added to the soil during the study are given in Fig. 3.

Methods of sample collection and analysis

5. Samples were collected from the dose tank, tile drains in the filter field, and background tile drains biweekly from November through May in 1987–1988 and 1988–1989. Samples were filtered through a 0.22-μm membrane filter and analyzed with standard methods for concentrations of NH$_4$-N, NO$_3$-N, Cl$^-$, and total soluble organic carbon (TOC). Electrical conductivity was also determined. We determined fecal coliform numbers on unfiltered samples using the membrane filter procedure.

Statistical methods

6. Statistical analysis included analysis of variance and mean separation to compare chemical and biological parameters in the filter field tile drains (T2, T3, and T4) and background tile drains (T7, T8, and T9).

7. Results and Discussion

Specific results and comparison to the literature add credibility to the study.

The septic tank effluent had mean TOC, NH$_4$-N, NO$_3$-N, and Cl$^-$ concentrations of 55, 41, <1, and 50 mg/L, respectively, and an electrical conductivity of 0.84 dS/m. The values are typical for septic tank effluent (Cogger and Carlile, 1984; and Piluk and Hao, 1989).

Specific results with reference to the figure.

Results indicated that 1 m of soil was highly effective in reducing TOC levels in the wastewater effluent. Samples collected from the tile drains in the filter field absorption bed showed a 96% reduction in TOC (Fig. 4). Background tile drains had significantly lower TOC levels than did tile drains in the filter field, and values were not significantly different between years.

332

8. The Cl^- concentrations decreased from 50 mg/L in the septic effluent to 35 mg/L in the tile drains from the absorption bed to 8 mg/L in the tile drains from the background area (Fig. 4). The electrical conductivity values showed a similar trend.

The NH_4-N did not move through the soil as indicated by levels of 1 and 0 mg/L in the drain tiles from the absorption bed and background areas, respectively (Fig. 4). Walker *et al.* (1973) reported that NH_4-N levels decreased to low levels within 20 cm of the filter field beds in four soils studied.

9. The mean NO_3-N level in the tile drains from the absorption bed was 4 mg/L and significantly greater than the 1 mg/L in the background drains (Fig. 4). Mean NO_3-N levels in the tile drains showed a significant increase from 1 mg/L in 1987–1988 to 5 mg/L in 1988–1989, whereas the Cl^- levels were not different between years.

The combination of NH_4-N retention by the soil exchange sites and denitrification of the NO_3-N resulted in a reduction of the $NH_4 + NO_3$-N/Cl^- ratio from 0.8 in the septic effluent to 0.1 in tile drain samples from the absorption beds. The reduced ratio would indicate that denitrification was occurring. Stewart and Reneau (1988) studied a low pressure distribution system in a soil with a seasonally fluctuating water table and reported active denitrification. It appears that nitrification occurs when the soil is aerobic, and denitrification takes place when the soil becomes saturated (Reneau, 1979).

Soil renovation of fecal coliforms was also demonstrated with geometric mean values in the tile drains in the absorption beds and background tile drains of 18 and 3/100 mL, respectively (Fig. 5). Reneau (1978) used tile drains in wet soils to study fecal coliform movement from septic systems and reported that bacterial numbers decreased with distance from the septic system, and the decrease was described by a logarithmic equation.

10. Conclusions

Analyses of water samples collected from the septic system dose tank, tile drains in the filter field absorption beds, and background tile drains showed that, in a poorly drained soil, a low pressure distribution system with tile drains resulted in a significant renovation of domestic wastewater within 1 m of soil. Future research will involve quantitating water volume in the tile drains to enable us to assess the NO_3-N loading rate (volume × concentration) of the septic system.

11. References

Cogger, C.G., and B.L. Carlile. 1984. Field performance of conventional and alternative septic systems in wet soils. J. Environ. Qual. 13:137–142.

Miller, D.W. 1984. Sources of ground-water pollution. p. 17–19. *In* Protecting ground water, the hidden resource. U.S. Environmental Protection Agency. Washington, D.C.

Piluk, R.J., and O.J. Hao. 1989. Evaluation of on-site waste disposal system for nitrogen reduction. J. Environ. Eng. Div. Am. Soc. Civ. Eng. 115:725–740.

Reed, B.E., M.R. Matsumoto, A. Wake, H. Iwamoto, and F. Takeda. 1989. Improvements in soil absorption trench design. J. Environ. Eng. Div. Am. Soc. Civ. Eng. 115:853–857.

Reneau, R.B., Jr. 1978. Influence of artificial drainage on penetration of coliform bacteria from septic tank effluents into wet, tile-drained soils. J. Environ. Qual.7:23–30.

Reneau, R.B., Jr. 1979. Changes in concentrations of selected chemical pollutants in wet, tile-drained soil systems as influenced by disposal of septic tank effluents. J. Environ. Qual. 8:189–196.

Reneau, R.B., Jr., C. Hagedorn, and M.J. Degen. 1989. Fate and transport of biological and inorganic contaminants from on-site disposal of domestic wastewater. J. Environ. Qual. 18:135–144.

Stewart, L.W., and R.B. Reneau, Jr. 1988. Shallowly placed, low pressure distribution system to treat domestic wastewater in soils with fluctuating high water tables. J. Environ. Qual. 17:499–504.

Walker, W.G., J. Bouma, D.R. Keeney, and F.R. Magdoff. 1973. Nitrogen transformations during subsurface disposal of septic tank effluent in sands: I. Soil transformations. J. Environ. Qual. 2:475–480.

Captions for figures

Fig. 1. Design of the low pressure distribution septic system with tile drains and the background tile drains used in the study.

Fig. 2. Cross section of an absorption bed and tile drain trenches.

Fig. 3. Monthly precipitation amounts during the 2-yr study.

Fig. 4. Mean chemical concentrations in the septic effluent, tile drains from the filter field, and background tile drains.

Fig. 5. Geometric mean fecal coliform numbers in tile drains from the filter field and for 1987–1989.

Captions for photographs

Sand-filled tile drains were installed between the absorption beds and in a background area.

A low pressure distribution system was installed in a Calloway soil.

Annotated Bibliography of Select References

Anholt, R. R. H. (1994). *Dazzle 'em with Style: The Art of Oral Scientific Presentation.* W. H. Freeman and Company, New York.

This book is conversational and easy to read, and Anholt gives practical, realistic advice on presentations. His suggestions on structuring the presentation and designing and using visual displays are based on the principles of simplicity and audience understanding. He provides excellent advice on preparing and delivering the presentation with voice control, body language, and enthusiasm. His comments on poster presentations and answering questions are also good.

Booth, V. (1993). *Communicating in Science: Writing a Scientific Paper and Speaking at Scientific Meetings*, 2nd ed. Cambridge University Press, Cambridge, UK.

Booth's succinct 78-page book on writing scientific papers and speaking at professional meetings is dedicated to T. W. Fline (Those Whose First Language Is Not English). He includes important conventions in grammar, punctuation, and other mechanics and describes manners, style, visual aids, and delivery techniques in speaking. He offers some suggestions on use and misuse of numbers, troublesome aspects of the language for those for whom English is a second language, and the preparation of figures and copy for print; in addition, he notes differences in British and American English.

Briscoe, M. H. (1996). *Preparing Scientific Illustrations: A Guide to Better Posters, Presentation, and Publications*, 2nd ed. Springer, New York.

Briscoe demonstrates good visual aids with discussion and many illustrations. Her book gives sound advice on all kinds of visual imagery used in scientific communications—drawings, photographs, graphs, tables. And she applies these forms to various media — journal manuscripts, presentations, posters. Her book is clear and helpful for all kinds of scientific communications.

Committee on the Conduct of Science. (1989). *On Being a Scientist.* National Academy of Sciences, National Academy Press, Washington, DC.

Directed toward students beginning their careers, this booklet recognizes the human element — human judgments, human values, human error, and human relationships — in scientific research. "Much of the first half of the booklet looks at several examples of the choices that scientists make in their work as individuals.... The second half deals largely with questions that arise during the interactions among scientists.... A final section touches upon the social context..." (p.1). The book views the largely unwritten code of professional conduct and personal integrity in science and provides some detail on questions concerning scientific values, human decisions, treatment of data, sources of error, fraud, plagiarism, appropriate credit, obligations to society, and other ethical issues.

Council of Biology Editors (CBE) Scientific Illustration Committee. (1988). *Illustrating Science: Standards for Publication.* CBE, Bethesda, MD.

This well-illustrated book presents perspectives on preparing and publishing artwork for science. Subject matter includes information on drawings and photographs of biological specimens; the design and drawing graphs and maps; and use of type styles and sizes, tone and color, halftones, line art, and other details in illustration processes. It also provides information on materials to use in producing illustrations, getting an illustration ready to print, and addressing legal and ethical considerations. Note that the Council of Biology Editors is now known as the Council of Science Editors.

Council of Biology Editors (CBE) Style Manual Committee. (1983). *CBE Style Manual,* 5th ed. CBE, Bethesda, MD.

This manual presents conventions of technical style for the sciences based upon standards established by both "international and U.S. organizations concerned with science or with information services" (p. xvii). It provides information on planning and writing the manuscript for publication, prose style (grammatical conventions), and editorial style in documentation and tables and figures. It offers guidelines for dealing with proofs, indexing, copyrighting, and ethical conduct in authoring and publishing a paper. It includes stylistic conventions used in special fields of plant sciences, microbiology, animal sciences, chemistry and biochemistry, and geography and geology. Note that the Council of Biology Editors is now known as the Council of Science Editors.

Council of Biology Editors (CBE) Style Manual Committee. (1994). *Scientific Style and Format: The CBE Manual for Authors, Editors, and Publishers*, 6th ed. Cambridge University Press, Cambridge, UK.

More comprehensive relative to stylistic details than the fifth edition, this manual contains conventions for all of the scientific disciplines and is a good attempt to simplify and encourage uniformity in publication style for all the sciences. In addition to conventions such as punctuation, abbreviations, capitalization, symbolization, and references, this comprehensive manual provides information on stylistic conventions unique to specific areas of science. It also provides information on formats in journals and books, including forms for tables, figures, and indices. It does not, however, contain any instructions for planning, writing, and submitting papers for scientific publication, such as those included in the fifth edition. (Watch for a new edition.) Note that the Council of Biology Editors is now known as the Council of Science Editors.

Day, R. A. (1998). *How to Write and Publish a Scientific Paper*, 5th ed. Oryx Press, Phoenix, AZ.

A "cookbook" on preparing manuscripts for publication in scientific journals, this book deals with all parts of the paper from title to references. It contains information on other issues in the publication process such as submission and review of the paper and short chapters on other kinds of scientific communication such as review papers, posters, and oral presentations. Day presents pleasant, even entertaining, discussion on the use and misuse of English and avoiding jargon; some information on what constitutes scientific writing; ethical and legal issues; and helpful appendices on matters of abbreviations and common word problems in scientific writing.

Dodd, J. S., ed. (1997). *The ACS Style Guide: A Manual for Authors and Editors*, 2nd ed. American Chemical Society, Washington, DC.

Similar to the fifth edition of the *CBE Style Manual*, this guide includes not only specifics for editorial style and conventions in chemistry publications but also suggestions by various authors on types of scientific communications, instructions on writing papers and making oral presentations, and side issues of what goes on in the publication process, as well as issues such as copyright and ethics. It is the authority for style in all areas of chemistry as the CBE manual is in the biological sciences.

Macrina, F. L. (2000). *Scientific Integrity: An Introductory Text with Cases*, 2nd ed. ASM Press, Washington, DC.

Macrina's text provides a background for discussing issues of integrity in science. For the graduate advisor, the chapter on mentoring is excellent, and for students and other researchers, the information on record keeping can be helpful. Discussions on integrity regarding authorship and peer review, conflicts of interest, the use of animals or humans in experimentation, genetic technology, and the ownership of data with patents and copyrights—all are presented well by Macrina and other authors of individual chapters and are accompanied by case studies. This text would be good for a class on professionalism and ethics in science.

Montgomery, S. L. (2003). *The Chicago Guide to Communicating Science*. University of Chicago Press, Chicago.

Montgomery's book is well-written and covers most of the kinds of scientific writing. It also provides brief but good information on oral presentations and helpful comments on using the Internet. I would particularly recommend it for someone who struggles with how to write well and has had little experience with conventions and expectations in scientific writing. For example, his chapters on proposals and on graphics are good introductions for someone who has never produced these. The book is easy to read and can help you develop the necessary attitude for success with scientific communications.

O'Connor, M. (1991). *Writing Successfully in Science*. HarperCollins Academic, London.

Most of this book is centered around writing for publication in scientific journals. O'Connor contends that "Journal articles constructed in this formal way have become the basic units of research publication and are the model for many other kinds of writing in science" (p. ix). She covers planning, writing, preparing data in tables and figures, revising, submitting, and checking proofs of such an article. She also includes information on preparing and making presentations with slides and posters. The book provides information on writing grant proposals, theses, and review articles and is good for writers whose first language is not English.

Peters, R. L. (1997). *Getting What You Came for: The Smart Student's Guide to Earning a Master's or Ph.D.*, revised ed. Farrar, Straus and Giroux, New York.

Peters speaks from his own experience and that of other graduate or former graduate students with whom he has associated on the problems and pitfalls that may accost the student pursuing a graduate degree. He is very candid in his discussion, not only about communication efforts but also everything in the graduate environment from choosing a school and an advisor to handling the stress to finally getting a job. A good book for an undergraduate to read before going to graduate school but also instructive for the graduate student.

Reif-Lehrer, L. (1995). *Grant Application Writer's Handbook.* **Jones and Bartlett, Sudbury, MA.**

Reif-Lehrer's is the best and most detailed advice I've found for writing grant proposals, especially those for the large funding agencies such as the National Institute of Health or the National Science Foundation. Her checklists on points to consider in proposals and her appendices on other resources available are equally valuable for writing proposals to industry, private foundations, or even graduate committees. She furnishes helpful examples, and her experience with the granting agencies themselves and a network of people involved is evident. (Watch for an updated edition.)

Smith, R. V. (1998). *Graduate Research: A Guide for Students in the Sciences,* **3rd ed. University of Washington Press, Seattle.**

Smith's book "is designed for self instruction" (p. ix). His text serves to orient students to ideas about graduate research programs by discussing the student's commitment to the program, the principles and ethics involved in scientific research, and planning and time management. He deals with practical matters of library research, writing, preparing theses, presenting and publishing papers, getting grant support, and even getting a job.

Stock, M. (1985). *A Practical Guide to Graduate Research.* **McGraw-Hill, New York.**

Stock's is a helpful handbook for the graduate student in science. It deals with the "realities of how research is actually conducted" and carries the student "through the graduate program and through development, execution, and completion of the research project" (p. vii). This completion requires oral and written communication. Stock presents a practical, realistic view of the graduate research program with advice from choosing a topic and an advisor to the job interview seminar. She includes ideas on writing grant proposals, theses, and journal articles and on presenting talks and visual aids.

Tichy, H. J., and Fourdrinier, S. (1988). *Effective Writing for Engineers, Managers, Scientists,* **2nd ed. John Wiley & Sons, New York.**

With numerous examples, this handy reference book for all kinds of writers provides a thorough discussion about organization and getting started writing.

It has a section on grammar, punctuation, and diction. And it discusses the literary prose style or character of writing that results from sentence structure, word choice, the arrangement of ideas, and other elements of construction. It is a valuable guide for organization and clarity in any kind of writing.

Zinsser, W. (1998). *On Writing Well: The Classic Guide to Writing Non-Fiction*, 6th ed. HarperCollins, New York.

Zinsser writes about the principles involved with any kind of writing, and the book itself serves as an outstanding example of writing well. These principles lead to "using the English language in a way that will achieve the greatest strength and the least clutter" (p. 6). His principles of readability include the audience, the diction, construction and grammatical usage. Then he gives forms to use in writing about people, places, and subjects such as science, business, or sports. He provides evidence that good writing has universal qualities not to be diluted by the discipline in which it communicates.

Index

Page numbers followed by *f* indicates figures; *t*, tables.

A

Abstract, 13, 112-114. *See also*
 Executive summary
 contents of, 113-114
 evolution of, 293-295
 for oral presentations, 176-177
 purposes of, 111, 114
 for research, grant proposals,
  55-58
 in sample manuscript, 255
 writing of, 24
Abstract, introduction, methods,
 results, discussion. *See*
 AIMRAD
Accuracy, 38
Acknowledgements, 197
ACS Style Guide (Dodd), 15, 303
Ad hoc committees, 206
Advisors, 70, 71-72
Agricultural Research Service, 219
AIMRAD (abstract, introduction,
 methods, results, discussion), 6
*American National Standard for
 Writing Abstracts*, 112
Anholt, R.R.H., 14, 147, 150,
 178, 187
Animation, 312, 314
ArticleFirst, 41
Attitude
 scientific communication and,
  8-9, 11-12
 towards self, 228

Audience
 avenues for, 219-220
 communication and, 7
 for group communication, 209
 job interview and, 144-145
 of nonscientists, 216-219
 oral presentations and, 175
 for poster presentations, 191,
  192-195
 question, answer session and,
  146-147
 subject, purpose and, 8, 220
 timing and, 178
 titles and, 290
 understanding of, 218-219
 visual aids and, 179
Auger, P., 137
Authorship, 128

B

Bar charts
 for data presentation, 116
 preparation of, 122-123
Bayles, M.D., 132
Beebe, S.A., 214
Beecher, H.W., 7
Beyond Culture (Hall), 240
Bibliography, 38, 96-98
Biographical information, 63

BIOSIS Guide to Abstracts, 112
Birch, J.W., 81
Blum, D., 222
Board meeting, 205-206
Body language, 157-158, 237-239
Bolker, J., 73, 81
Booth, V., 14, 140, 150, 186, 187
Brainstorming, 23, 207
Briscoe, M.H., 14, 15, 121, 123, 150,
     156, 194, 201, 204
Budget, 62
Burnett, R.E., 14, 221
Buzz session. *See* Round-table
     discussion

C
Campbell, S., 17, 41, 45
Carlyle, Thomas, 133
Cartographic Relief Presentation
     (Imhof), 305
Cause-effect, 32
CBE. *See* Council of Biology
     Editors
CBE Style Manual, 86, 88, 108, 112,
     114, 119, 128, 135, 303
Charts, 121
Chilberg, J.C., 214
Closed panel, 212-213
Color
     area/size relationship of, 308
     combination of, 306-308
     communication with, 154-156
     composition of, 306, 308-310
     conventions, expectations
          of, 154
     importance of, 155-156
     intensity of, 308
     for poster presentations, 197-199

Color *(Continued)*
     in slide, poster presentations,
          154-156
     subdued v. pure, 307
     in visual aids, 305-310
Colton, C.C., 99
Committee meeting, 205-206
Committee on Science,
     Engineering, and Public
     Policy, 126
Committee on Sources, Dartmouth
     College, 127
Communication, scientific
     in American English, 231
     attitude and, 8-9, 11-12
     audience for, 7
     avenues for, 219-220
     biases and, 216
     body language and, 157-158
     changes in, 15-16
     clarity, simplicity of, 7
     clear, 1-2
     color and, 154-156
     cultural, environmental
          background and, 227-228
     direct v. indirect, 228-229
     dishonesty in, 127
     ethical, legal concerns of, 125
     ethics in, 125-128
     in group, 205-223
     human element of, 17
     importance of, 1
     informal, 141
     kinds of, 13
     listening and, 159-161
     literature searches and, 36-37
     for media, 219-220
     to nonscientist, 215-216
     nonverbal, 151-161
     oral, visual media of, 222
     organization, content of, 5-6

Communication, scientific
(*Continued*)
  patents and, 136
  physical, 156-159
  position, posture, space in,
    158-159
  poster presentation and, 192
  power distance and, 229-230
  preparation for, 11
  purpose of, 5
  questions and, 5-6
  respect for, 12
  reviewing, revising and, 103
  semantic environment of, 3-4
  setting and, 156
  simplicity, clarity in, 12-13,
    18, 153
  skills for, 4
  slides and, 165-173
  speaking and, 4
  style for, 15
  subject of, 8, 220
  subjective, 3
  symbolic, 144
  symbols and, 152-153
  tangible details and, 33
  techniques for, 6
  technology and, 16-17
  type style in, 153-154
  understanding of, 217-218
  visual aids to, 163-173
  writing, presentation of,
    222-223
Comparison-Contrast, 32, 33
Conclusion slides, 181, 188
Conclusion(s)
  in oral presentation, 183
  organizational format and, 29
  in poster presentation, 197, 332
  in proposal, 61
  in sample manuscript, 258-259

Conclusion(s) (*Continued*)
  structure of, 232
  in thesis, 74
Confidentiality, 130
Content
  of communication, 5-6
  inappropriate, 244-245
  revisions to, 105
Cooperative Extension Service, 219
Copyright, 133-135
  duration of, 134
  of electronic communication, 134
  "fair use" v. commercial use
    and, 134
  granting permission of, 134
  patents v., 135-136
  permission request for, 135,
    303-304
  registration of, 134
Corneis, Cary, 328
Council of Biology Editors (CBE),
    15, 94, 156
Council of Graduate Schools,
    76, 77
Cousteau, Jacques, 222
Cover sheet, 56*f*
Crawford, S.Y., 92
Crazy Talk, Stupid Talk
    (Postman), 3
CRIS, 41
Criticism, 285-288
Cultural differences, 225-226
  of attitudes toward self, 228
  direct v. indirect communication
    and, 228-229
  eye contact and, 238-239
  plagiarism and, 233-234
  power distance and, 229-230
  in speaking, 234-235
  titles and, 236-237
  towards time, 230

Culture(s)
adjustment to, 226-228
differences in, 225-226, 228-230
direct, indirect communication
and, 228-229
power distance and, 229-230
self and, 228
Current Contents Search, 41
Current search, 36
Curriculum vitae, 63

D

da Vinci, Leonardo, 307
The Dance of Life (Hall & Hall), 240
Data
clarity of, 116-117
distortion of, 126
ethics and, 129-130
figures, tables and, 115, 120-123,
297-301
graphic standards for, 121
honesty and, 127
presentation of, 115-123, 200-201
representative, 115
simplicity of, 121
visual display standards for, 121
writing errors and, 245
Davis, E.B., 15, 39, 41
Day, R.A., 14, 28, 46, 87, 92, 105,
112, 114, 119, 222, 233
DeBakey, L., 52, 53, 62, 65
Definition, 32, 33
Department of Defense, 45
Departmental seminars. *See*
Seminars, departmental
Descriptive abstract, 112-113.
See also Abstract
Diction, 248

Discussion
leader of, 209-211
in oral presentation, 183
organizational format and, 29
in poster presentation, 197,
331-332
in sample manuscript, 258-259
slide presentation and, 188
in thesis, 74
Discussion groups, 212-214. *See also*
Group communications
Discussion leader
preparation by, 209-210
responsibilities of, 210-211
Dissertation, 13, 76. *See also*
Thesis(es)
journal manuscripts and, 78-81,
80-81
Dissertation Services, 75
Documentation. *See also* References
accuracy, style in, 96-99
of electronic sources, 98
of research literature, 37
Dodd, J.S., 108, 135
Dreyfuss, H., 152, 154, 155

E

Editing, 103-105
Ehrlich, Paul, 327
EndNote, 37
Enumerate, 32
Errors, 247
Ethics, 125
collaborations and, 132
confidentiality and, 130
conflicts of interest and, 131
crediting others and, 128
errors of, 126

Ethics *(Continued)*
  group communications and, 207
  mistakes and, 130
  omission and, 132
  others' time and, 131
  professional respect and, 128-129
  in reporting data, 129-130
  republishing and, 130
  standards of, 132-133
  time, effort and, 131-132
  in workplace, 131
Executive summary. *See also*
    Abstract
  importance of, 56
  for research, grant proposals,
    55-58
Exploratory search, 35-36
Extended abstract, 112.
    See also Abstract

F
*Faculty and Student Challenges
  in Facing Cultural and
  Linguistic Diversity* (Clark &
  Waltzman), 240
A Field Guide for Science Writers
  (Blum & Knudson), 222
Figures
  for data presentation, 120-123,
    298-301
  forms of, 120
  preparation of, 122
  in slide presentation, 169
Fitzpatrick, J., 81
Folt, C.L., 59
Font, 153-154
Format
  of literature review, 47

Format *(Continued)*
  for thesis, 76, 281-282
  for writing, 26-28
Forum, 209, 212
Fourdrinier, S., 14, 28, 105
Fox, H., 230, 233, 240
Friedland, A.J., 59

G
Gaffney, N.A., 77
Gannon, R., 222
Gastel, B., 115, 219, 221, 222
Gentry, Terry, 49
Getting What You Came For
  (Peters), 13, 71, 240
Gilpin, A.A., 14
Gordon, C.H., 127
Gould, S.J., 126, 222
Graduate program
  requirements for, 69
  thesis and, 67
Graduate proposal, 13
  audience for, 52
  budget, time frame for, 63
  purpose of, 51-52
Grant proposal, 13
  audience for, 52
  biographical information for, 63
  budget, time frame and,
    62-63
  preparation for, 53
  purpose of, 51, 52
  rejection of, 64
Graphics. *See also* Figures;
    Illustrations; Tables
  data presentation with,
    120-123
  standards for, 121

Graphs
  color, size in, 200-201
  in posters, 200
  preparation of, 122
Group communications
  with audience, 209-214
  avenues for, 219-220
  brainstorming and, 207
  closed panel and, 212-213
  cooperation and, 208
  decision making in, 206
  discussion leader in, 209-211
  ethics in, 207
  forum and, 209, 212
  groupthink in, 208
  important points for, 213
  media and, 219-220
  member responsibilities for,
    211-212
  panel discussion and, 209
  principles of, 214
  problem solving in, 206-207
  simplicity in, 221
  size and, 207
  structure of, 214
  subject of, 220
  symposium and, 209
  techniques for, 221-222
  without audience, 205-209
Group members, 211-212
Groupthink, 208

H
Haakenson, R., 147
Hall, E.T., 151, 226, 228, 237, 240
Hall, M.R., 151, 237, 240
A Handbook for Scholars
  (van Leunen), 14

Handouts, 202
Hanna, M.S., 214
Hays, Laura, 155
Headings, 95, 152
Headline News/Science Views
  (Jarmul), 222
Henderson, J., 44, 45
Hess, Harry, 328
Hodges, E.R.S., 121, 123
How to Write and Publish a
  Scientific Paper (Day), 92
Hurd, J.M., 92
Hypothesis
  null, 58-59
  proposal introduction and, 59

I
Illustrating Science (CBE), 15, 121,
  123
Illustrations, 15
Imhof, E., 123, 154, 155, 156, 161,
  192, 197, 202, 233, 305
IMRAD (introduction, methods,
  results, discussion), 6, 26, 28
Indicative abstract. See Descriptive
  abstract
Informative abstract, 112, 113
Ingenta, 41
International students
  advice for, 239
  body language for, 237-239
  casual conversation for,
    236-237
  cultural adjustment for,
    226-228
  cultural differences and, 225
  eye contact and, 238-239
  language and, 225-226

Introduction
  in oral presentation, 182
  for poster presentation, 195, 330
  purpose of, 28
  for research, grant proposals,
    58-59
  in sample manuscript, 255-256
  in slide presentation, 187
  in standard American English,
    231-232
  to thesis, 74
Introduction, methods, results,
  discussion. *See* IMRAD

J
Jacobs, P.F., 15, 39, 41
Janis, I.L., 208
Jarmul, D., 222
Jensen, A.D., 214
Job interview
  audience for, 144-145
  purpose of, 144, 145
  responses to, 146
  speaking at, 144-146
  suggestions for, 145-146
Journal, scientific
  copyright permission for, 135
  reviews, 107
Journal articles, 13
  documentation style for,
    96-99
  heading style for, 95
  technical style for, 93-95
Journal manuscripts, 77
Journal reviews, 107
Journal(s), scientific
  contributions to, 85
  editing, reviewing of, 89-92

Journal(s), scientific *(Continued)*
  planning, writing paper for,
    86-87
Justification, 59-60

K
Keyes, E., 153
Knisely, K., 14, 39, 150, 204
Knudson, M., 222

L
Language
  bias, 44
  changes in, 15-16
  cultural differences and, 234-235
  differences in, 12
  variations in, 48-49
Layout, 199-200
Legal issues, 133-135
Lehr, Jay, 142, 150, 175, 323
"Let There Be Stoning" (Lehr), 175,
  323-328
Levy, R.C., 137
Lewis, Ricki, 222
Library
  for literature searches, 39-40
  research literature and, 36
  thesis and, 70
Line graphs, 116, 123
List, C., 41
Listening
  importance of, 159
  objective of, 159-160
  in small group, 160
  tips for, 160-161

Listening to the World (Fox), 240
Literature
  documentation of, 37
  evaluation of, 42-46
  information sources for, 39-41
  relevance, credibility of, 43
  review of, 46-49, 60-61
  search for, 35-39
  selection of, 37-38, 42
  sources for, 42-46
  verification of, 38-39
Literature cited, 74. *See also*
  References
Literature review. *See also*
  Literature
  elements of, 60-61
  organization of, 261
  problems with, 48
  purpose of, 46
  rough draft of, 74, 281
  sample of, 261-268
  suggestions for, 47-49
  writing of, 46-47
Luchsinger, A.E., 119
Luellen, W.R., 14-15, 101, 204
Lutz, J.A., 222

M
MacGregor, A.J., 121
Macrina, F.L., 72, 128, 132, 137
Manuscript. *See also* Paper,
  scientific
  abstract in, 255
  conclusions in, 259
  introduction in, 255-256
  materials, methods in, 256-257
  organization, development of,
    253-259

Manuscript *(Continued)*
  results, discussion in, 258-259
  reviewing of, 108-109
  revisions to, 106-107, 285-288
  sample of, 253-259
Marshek, K.M., 31
Masterson, J.T., 214
Materials and methods
  in poster presentation,
    195, 331
  purpose of, 28
  in sample manuscript, 256-257
  slide presentation and, 188
  in thesis, 74
Mauch, J.E., 81
Maxwell, John, 328
Mays, Thomas D., 137
McCown, B.H., 178
McMillan, V.E., 14, 39, 113
Media, 219-220
Meetings, professional
  getting most from, 142
  informal communication
    at, 141
  preparation for, 142
  presentations at, 142-144
Methods
  oral presentation of, 182-183
  in proposal, 61
Microsoft Power Point, 166,
    167, 173
The Mismeasure of Man
  (Gould), 126
Moderator
  role of, 148-149
  suggestions for, 327
  tips for, 149
  transitions and, 184, 185
Montgomery, S.L., 14, 28, 240
Morris, Justin R., 293
Munger, D., 17, 41, 45

N

National Association of Science Writers, 222
Nicholson, Richard S., 223
Null hypothesis, 58-59

O

Objective
    isolation of, 37
    in poster presentation, 195, 331
    proposal introduction and, 59
Objective slides, 181
O'Connor, M., 14, 65, 81, 87, 92, 192, 194, 197, 204
On Being a Scientist, 13, 86, 128
On Writing Well (Zinsser), 14
Oral presentation. *See also* Presentations; Speaking
    abstract for, 176-177
    conditioning for, 176
    content for, 180*t*
    influences on, 234-236
    for international students, 234-236
    introduction for, 182
    at meetings, 323-328
    methods portion of, 182-183
    microphone for, 177-178
    peer review of, 187
    results, discussion portion of, 183
    timing for, 178-179
    tips for, 177
    transitions in, 183
    visual aids and, 179
Organization
    of communication, 5-6
    coordination of, 28-29
    framework for, 26-27

Organization *(Continued)*
    generic outline for, 244
    of literature review, 47-49
    of rough draft, 34
    sections for, 28-29
    of slide presentation, 181-182
    thesis and, 29-30
    for writing, 26-28
Outline, 23-24
    for literature review, 47-48
    organizational methods and, 27-28
    in sample manuscript, 253-254
    for traditional thesis, 78

P

Pachet-Golubev, P., 14
Panel discussion, 209
Paper, scientific. *See also* Manuscript
    acceptance of, 89
    body of, 232
    journal process for, 89-92
    planning, writing of, 86-87
    proofreading of, 100-101
    rejection of, 90-91
    reviewers' comments for, 106-107
    reviews of, 87-88
    revisions to, 103, 104-105
    structure of, 231-233
    style of, 93-95
    submission of, 88-89
Paradis, J.G., 105
Patents
    communications and, 136
    copyright v., 135-136
    criteria for, 136
    literature of, 136

Patents *(Continued)*
resources for, 137
use of, 135-136
Paul, J.K., 136
Peer review, 107, 187
Permission, 303
Peters, R.L., 13, 65, 73, 81, 83,
150, 240
Photography
in posters, 201
in visual presentations, 167
Pie charts, 116
Plagiarism, 127-128, 233-234
Plotnik, A., 166
Poster presentations. *See also*
Presentations
acknowledgements in, 197
advantages of, 192
audience for, 191, 192-194
color, physical qualities of,
197-199
content, organization of, 195
handouts for, 202
as "illustrated abstract," 194
presenter and, 201-202
space, simplicity of, 202-203
spacing and, 196*f*
text and, 194-197
tips for, 194
type size, style and, 197
understanding of, 194-195
use of, 191
Poster(s), 13
characteristics of, requirements
for, 143*t*
color in, 305
oral presentation and, 178
photographs in, 201
presenter of, 201-202
recommendations for, 204
sample layout for, 329*f*

Poster(s) *(Continued)*
sample text for, 329-333
spacing, arrangement in, 199-200
tables for, 120
time, construction of, 203-204
type sizes for, 198*t*
Postman, Neil, 3
Power distance
communication and, 229-230
cultural differences of, 229-230
eye contact and, 238-239
issues of, 236
physical distance and, 238
*A Practical Guide to Graduate
Research*, 67
Presentations
audience and, 7, 142-143
backup plan for, 149
central idea for, 143-144
characteristics, comparisons
of, 143*t*
color in, 154-156
communication ability and, 141
content for, 180*t*
of data, 169
at departmental seminars,
140-141
help for, 14
importance of, 139-140
job interview and, 144-146
microphone for, 177-178
moderator at, 148-149
new perspectives and, 140-141
oral, 175-189
organization, content of, 6
poster v. slide, 192
preparation for, 142
at professional meetings, 142-144
question, answer session at,
146-148
sample table for, 170*t*, 171*t*

Presentations *(Continued)*
  slide composition and, 165-172
  suggestions for, 150, 323-328
  transitions in, 183-187
  type style for, 153-154
  types of, 13
  visual aids for, 144, 163-173
Problem solving
  in group communication, 206-207
  procedure for, 206-207
ProCite, 37
Progress reports, 64-65
Proofreading
  marks for, 100
  of scientific paper, 99, 101
Proposal(s)
  abstract for, 55-58
  conclusions in, 61
  considerations for, 63-64
  content, form of, 53-55
  executive summary for, 55-58
  format for, 26
  for graduate thesis, 72
  guidelines for, 53
  help for, 13-14
  introduction for, 58-59
  judging criteria for, 54
  justification for, 59-60
  methods section of, 61
  organization for, 54-55
  parts of, 54
  preliminary work and, 269
  reference section for, 61-62
  sample of, 269-278
  title, title page for, 55
  types of, 13, 51
Publication
  copyright and, 133
  documentation style for, 96-99
  ethics of, 130
  manuscript review for, 285-288

Publication *(Continued)*
  process of, 89-92
  resources for, 92
  of scientific paper, 87
  of thesis, 75
Purpose
  audience and, 8
  of communication, 5

Q
Question and answer session
  in formal presentation, 146-148
  preparation for, 147
  suggestions for, 147-148

R
Ransom, Nora, 13, 24
References. *See also* Literature cited
  accuracy, style in, 96-99
  in poster presentation, 197,
    332-333
  for proposals, 61-62
RefWorks, 37
Reid, W.M., 119
Reif-Lehrer, L., 52, 53
Reports, 13
  audience for, 7
  help for, 13-14
  organization, content of, 6,
    231-233
Representative data, 115
Research, scientific
  electronic sources and, 44-45
  language bias in, 44
  prejudices and, 126

Research, scientific *(Continued)*
  source credibility, reliability in,
    42-45
  source selection, evaluation for,
    42-46
  topic for, 281
  writing, speaking about, 2
Research projects, 64-65
Resources
  for literature searches, 39-41
  for publishing papers, 92
Results
  data presentation and, 116
  in oral presentation, 183
  organizational format and, 29
  in poster presentation, 195, 197,
    331-332
  in sample manuscript, 258-259
  slide presentation and, 188
  thesis and, 74
Reviewers
  role of, 108
  suggestions from, 106-107
Reviewing
  of others' work, 107-109
  of own paper, 104-105
  principles for, 108
  suggestions for, 108-109
Reviews, 107
Revisions
  to own work, 104-105
  reviewers' comments and,
    106-107
Rogers, Will, 215
Rough draft
  approaches to, 31-34
  checklist for, 34
  of literature review, 74, 281
  organization, development
    and, 250
  of results, 282

Rough draft *(Continued)*
  revision of, 250
  sample of, 249-251
  starting of, 22
  writing of, 31-34
Round-table discussion
  (buzz session), 205
Running head, 112
Running title, 112

S
Sagan, Carl, 222, 327
Samuelson, Paul, 327
Schmid, C.F., 121
Schmidt, D., 15, 39, 41
Science Illustration Committee, 117
Scientific communication.
    See Communication, scientific
Scientific English (Day), 14
Scientific Style and Format (CBE),
    15, 69, 94
Scientific writing. *See* Writing
Search(es)
  accuracy for, 38
  approaches to, 40-41
  documentation of, 37
  guides for, 41
  objectives for, 37
  selectivity of, 37-38
  snowball, 40-41
  types of, 35-36
Secrist, J., 81
Semantic environment, 3
Seminars, departmental
  opportunities at, 140-141
  presentations at, 140-141
  research evaluations and, 141
Setting, 156

Silyn-Roberts, H., 6, 14, 48, 65, 204
Simmons, P., 127
Slang, 237
Slide presentation, 13. *See also*
  Presentations
  animation in, 312, 314
  balance in, 167
  characteristics of, requirements
    for, 143*t*
  check list for, 187-189
  composition of, 165-172
  conclusion for, 181, 186, 188
  criteria for, 166
  introduction for, 187
  materials, methods checklist
    for, 188
  objective slides for, 181
  organization of, 181-182
  photography and, 167
  poster presentation v., 192
  recommendations for, 167-169
  results, discussion checklist
    for, 188
  sample of, 315-322
  suggestions for, 171-172, 326
  tables for, 120
  title slides and, 181
  transitions in, 183-187
  typical content for, 180*t*
  visual aids for, 188
Slides
  animation in, 312, 314
  benefits of, 325-326
  color in, 305
  composition of, 165-172, 173
  production of, 172-173
  simplicity in, 173, 311-312, 314
Smith, R.C., 119
Smith, R.V., 13, 39, 81
Smith, T.C., 157, 221
Snowball technique, 40-41

Sources
  credibility, reliability of, 42-45
  documentation of, 46
  electronic, 44-45
  evaluation of, 42-46
  selection of, 42
  unpublished references as, 46
Speaking. *See also* Oral presentation
  casual conversation, 236-237
  direct v. indirect communication
    and, 229
  for international students, 234-236
  speed, pace, volume and, 235-236
Specific search, 36
Speech. *See* Oral presentation
Standing committees, 206
Stock, M., 59, 67, 81
Storms, C.G., 222
Structure
  data and, 245
  of sentences, 245-247
  of writing, 245-247
Style
  for headings, 95
  technical, 93
Style sheets, 69-70
Subject, 8
Symbol Sourcebook (Dreyfuss), 152
Symbols
  to clarify, organize, 152-153
  for communication, 152-153
  conventions for, 152
Symposium, 209, 213

T

Tables
  for data presentation, 116, 297-298
  parts of, 118-119

Tables *(Continued)*
  in posters, 200
  presentation sample of,
    170*t*, 171*t*
  problems with, 118
  publication guidelines for,
    119-120
  in slide presentation, 169
  use of, 117-118, 120-121
Tangible details, 33
Task force, 206
Technical Communication
  (Burnett), 14
Technology
  clarity and, 18
  communication and, 16-17
  drawbacks of, 16-17
Tenner, Edward, 16, 154, 155
Text, 195
Thesis defense, 282-283
  form, style of, 83
  questions at, 83
Thesis(es), 13
  advisors for, 70, 71-72
  alternate routes to, 280*f*
  approval of, 75-76
  avoiding problems in, 70-71
  content of, 68-69
  data for, 72-73
  defense of, 82-83, 282-283
  differences in, 68
  final chores for, 75-76
  final copy of, 74
  finishing of, 71, 74, 283
  format of, 281-282
  formats for, 77-82
  foundations for, 68
  help for, 13-14
  hurdles for, 279-283
  journal manuscripts and,
    78-82

Thesis(es) *(Continued)*
  library and, 70
  masters, 79-80
  organization of, 29-30
  planning for, 76-77
  proposal for, 72
  publishing of, 75
  requirements for, 69-70
  resources for, 69-70
  revisions to, 282
  specialists for, 70
  structure of, 231-233
  style sheets for, 69-70
  time line for, 67
  topic for, 281
  traditional, 78
  as work in progress, 73-74
Tichy, H.J., 14, 28, 105
Tierney, E.P., 150
Time, 230
Time frame, 62
Title page, 55
Title slide, 181
Title(s), 111-112
  active verb in, 290
  audience and, 290
  evolution of, 289-291
  problems with, 112
  purposes of, 111
  words in, 289
Topic, 281
Transitional device, 30-31, 184
Transitions
  devices for, 30-31, 184
  in slide presentations,
    183-187
Tufte, E.R., 117, 121, 156
Type
  in poster presentation, 197
  sizes of, 198*t*
  styles of, 153-154

U

Understanding Cultural Differences (Hall & Hall), 240
U.S. Copyright Office, 133, 134
U.S. cultures, 226-228
U.S. Department of Agriculture, 45
U.S. Patent and Trademark Office, 136

V

Vander Stoep, Gail, 233, 285, 303
van Kammen, D.P., 60
van Leunen, M-C., 14
Venolia, J., 14
Visual aids
  color combination in, 306-308
  color composition in, 306
  color use in, 305-310
  criteria for, 166
  design rules of, 309-310
  function of, 163
  look of, 165
  for oral presentation, 179
  photography and, 167
  for slide presentation, 188
  speech and, 179-183
  tips for, 163
  types of, 164
  values, limitations of, 164-165

W

The Wall Street Journal, 154
Weller, A.C., 92
Wherry, T.L., 137
Why Things Bite Back (Tenner), 16

Wilson, G.L., 214
Wolfgang, Pauli, 244
Woolever, K.R., 240
Woolsey, J.D., 14, 150, 193, 195, 199, 200, 204
Words
  audience and, 7
  semantic environment of, 3, 7
  understanding of, 2
Wright, D.J., 81
Write Right (Venolia), 14
Writing
  approaches to, 32
  cultural differences in, 230-231, 235
  diction in, 248
  direct, simple approach to, 230-231, 233
  direct v. indirect communication and, 229
  help for, 14
  inappropriate content of, 244-245
  inconsistencies, errors in, 247
  for international students, 230-234
  of literature review, 46-49
  organization, development of, 26-28, 28-29
  plagiarism and, 127
  preparedness for, 243-244
  prewriting exercises for, 22-26
  questions for, 22-23
  rough draft and, 31, 34
  rules for, 25
  science v. scientific, 222
  scientific, technical, 13
  for scientific journals, 85-87
  starting of, 22
  structure, unity of, 245-247
  thinking and, 22
  transitions in, 30-31
  weaknesses in, 243

Writing for the Technical Professions
  (Woolever), 240
Writing Scientific Papers in English
  (O'Connor & Woodford), 92
Writing Successfully in Science
  (O'Connor), 92
Writing to Learn (Zinsser), 222
Wynn, G., 127

Z
Zimmerman, M.L., 105
Zinsser, W., 14, 22, 103, 105, 111,
  221, 222